SPRINGER
LABORATORY

G. R. Newman   J. A. Hobot

Resin Microscopy and On-Section Immunocytochemistry

With 25 Figures

Springer-Verlag
Berlin Heidelberg New York London Paris
Tokyo Hong Kong Barcelona Budapest

Geoffrey R. Newman, Ph. D.
Jan A. Hobot, Ph. D.

Electron Microscopy Unit
University of Wales College of Medicine
Heath Park
Cardiff CF4 4XN
Great Britain

ISBN 3-540-56429-2 Springer-Verlag Berlin Heidelberg New York
ISBN 0-387-56429-2 Springer-Verlag New York Berlin Heidelberg

© Springer-Verlag Berlin Heidelberg 1993
Printed in Germany

Typesetting: Camera ready by author
2217/3145-5 4 3 2 1 0 – Printed on acid-free paper

Foreword

At the outset it is pointed out that the present text is written by two prominent, actively practicing scientists. They have published a large number of excellent articles and chapters in the fields of histochemistry, immunohistochemistry, cytochemistry, and immunocytochemistry. I have greatly benefited by reading their previous publications.

The book is divided into two major topics: RESIN EMBEDDING and ON-SECTION IMMUNOLABELLING. The importance of resin sections in the aforementioned fields cannot be overemphasized. Although fluorescence microscopy has played a key role in determining the localization and structure of cellular components it provides a relatively low resolving power. Thin resin sections, on the other hand, have the advantage of yielding images of the ultrastructure at high resolutions. For this and other reasons the use of thin resin sections has become the most important appproach for the *in situ* localization of cellular structures including antigens with the electron microscope.

The presentation of both theory and practical aspects of the methodology is one of the most important and useful contributions of the book. The acceptance of a preparatory procedure without understanding the theory or principle is justified only on the basis of blind faith. The comprehensiveness of this publication is reflected by the fact that the discussion of both the classical and modern techniques is included.

It is not uncommon that the ability to complete a preparatory procedure depends upon the knowledge of *all* the details of the methodology. This necessity is fulfilled by the present book, for it contains the most comprehensive details of the methods. The authors have made certain that even the smallest details of a protocol are presented. This approach will result in carrying out the procedures accurately and successfully. This publication is expected to be eagerly welcomed by a large number of novice as well as experienced research workers.

Kean College of New Jersey,   M.A. Hayat,
                                                               Professor.

Preface

The introduction of antibodies tagged with markers and used to identify tissue substances (Coons et al, 1941; Nakane and Pierce, 1966; Sternberger, 1979) has led to the development of the science of immunocytochemistry but some basic questions of how best to prepare the tissue still remain. Resin embedded tissue is now routinely used for immunomicroscopy techniques, although frozen sections and paraffin wax embedded tissue still dominate light microscope immunocytochemistry. Nonetheless, new techniques and approaches are constantly introduced, and the novice entering into this field has a breathtaking variety of methods open to him. Even microscopists find it difficult to keep up to date with the latest innovations. Books, with chapters contributed by experts, have appeared, but the reader is often left without a sense of order or perspective from which to formulate the best way to start a working protocol to fit a particular problem. We have tried to overcome this by presenting an overall strategy into which the various techniques available for resin embedding are logically introduced.

The resins that have provided the most excellent results for immunomicroscopy are the modern commercially available acrylics (LR resins; Lowicryls). Epoxides, though, cannot be discounted. Many laboratories have material embedded in these resins for which limited immunocytochemistry is still a possibility. Therefore, methods involving the epoxides are included, even though the net result is an end-product that is less sensitive to immunotechniques. The strategy, upon which this book is based, covers the embedding of tissue using less sensitive epoxy resin methods to the more sensitive procedures either at room temperature or low temperature employing the modern acrylics. That this is possible is discussed and results presented to the reader so that an understanding of the techniques can be acquired and appropriate choices made.

However, we do not wish solely to describe methods that, although successful, require a tremendous expense in time, equipment and attention to experimental procedure. A great deal of work can be done very simply and cheaply with high levels of immunosensitivity and good ultrastructure! We discuss, in the first part of the book, the background of the various resins available and provide information on inexpensive alternative technologies where they exist. The various steps involved in tissue processing, beginning with fixation, are first described in theory, then detailed protocols are presented for their applications. Areas where problems can arise are included in this treatment of the protocols. Further, throughout the book extensive cross-referencing to original studies and their results is included. The references listed will therefore allow readers to widen their knowledge of particular areas of interest.

The second part of the book rationalises the great variety of labelling methods that are commonly used for "on-section" cytochemistry and immunocytochemistry. Colloidal gold, in a variety of sizes, can be purchased linked to a bewildering number of primary and secondary detection substances, for example enzymes, immunoglobulins, lectins, protein A, protein G, avidin and so on. In addition to transmission electron microscopy (TEM), it can be used for light microscopy (LM) and scanning electron microscopy (SEM). The accuracy with which colloidal gold particles can be sized means that different sizes can be used for double labelling.

Enzyme markers were the first to be adapted for EM use. Of these, peroxidase/diaminobenzidine (DAB) has stood the test of time and is the most frequently used. Direct, indirect, bridge and sandwich techniques, dependant for the accuracy of their localisation on either natural determinants like PAP (Sternberger, 1979) or artificial haptenoids such as biotin, still have a part to play in modern electron microscopy, despite the present overwhelming popularity of colloidal gold. Peroxidase/DAB can also be combined with colloidal gold for double labelling. The principles behind their usage and their limitations and advantages are discussed.

Finally, there are protocols for specific labelling techniques applied to semithin and thin sections of different resins and detailing their different requirements, before, during and after labelling, with appropriate references to trouble shooting.

University of Wales College of Medicine,   Geoffrey R. Newman
Cardiff, April 1993.   Jan A. Hobot

Acknowledgements

We would like to express our heartfelt thanks to Dr. Bharat Jasani (University of Wales College of Medicine, Cardiff, UK) for his invaluable comments and advice during the writing of this book. We also extend our appreciation to Drs. Audrey Glauert and Peter Lewis (University of Cambridge, UK) for helpful discussions and for critical reading of the manuscript. And finally, a special word of gratitude must be extended to Mrs. Sim Singhrao and Mrs. Val James of the EM Unit at the UWCM, Cardiff, for their outstanding patience and help whilst the authors were engaged in writing.

Table of Contents

PART I: RESIN EMBEDDING

1   The Strategic Approach

1.1   Introduction

In biological electron microscopy (EM), the last fifty years have been largely dedicated to the elucidation of cellular ultrastructural detail. The developmental research that has been needed to resolve cell structure is now part of the history of the subject. Fixation and resin embedding methods for observing structure have been relegated to the routine in most laboratories. The ultrastructural aim is necessarily towards stabilising tissue to protect its structure during the deleterious process of embedding. Embedding tissue in a powerfully crosslinked resin makes it possible to produce the incredibly thin yet strong sections that resist damage caused by an electron beam and provide high resolution. Since the work of Sabatini, Bensch and Barrnett (1963) double fixation with neutral buffered glutaraldehyde and osmium tetroxide has been confirmed to conveniently preserve the finest ultrastructural detail. Dehydration in an organic solvent gradient is then followed by embedding in an epoxy resin. Araldite, for example, is as popular now as it was shortly after the first description of its use by Glauert et al (1956).

However, electron microscopists of today are at least as interested in the localisation of cellular substances as they are in the identification of cell structures and this has resulted in a resurgence of interest in methodology. By definition, tissue which has been chemically altered (stabilised) by fixation and epoxide embedding cannot be highly sensitive to cytochemical, histochemical, immunocytochemical and molecular biological techniques, with the result that methods that restrict or even eliminate chemical fixation are gaining prominence. Inevitably, these methods result in tissue which is vulnerable to damage during dehydration, infiltration and embedding so that revised tissue processing regimes have also been required. Finally, the tough, unyielding, three-dimensionally bonded, hydrophobic epoxides, responsible for so much of our understanding of tissue structure, are giving way to gentler, more versatile, hydrophilic, electron beam stable acrylic resins better suited to methods aimed at retaining antigenic reactivity within tissue.

In this context, the recent evolution of a number of well-defined principles has led to the resin embedding methods detailed in this book (Fig. 2, Sect. 1.3.1). For example, the electron microscopical routine, based upon epoxy embedding procedures and dedicated to ultrastructure, often leads to very low sensitivity on-section immunolabelling. However, immunolabelling of more robust antigens in some tissue sites is still possible. Insulin in pancreatic islet cells and amylase in pancreatic acinar cells are good examples of antigens that are still recognisable to their antibodies after post-osmication and epoxide embedding. In addition, many electron microscope laboratories have a back-log of osmicated epoxide embedded tissues which could be very interesting to re-investigate by immunocytochemistry. For these reasons any descriptions of methodology must begin with the traditional glutaraldehyde/osmium/epoxide method.

There is a wealth of information on this method so it is only briefly described in this book. Instead more space is devoted to epoxide section pretreatment which can sometimes restore antigenic sensitivity within tissue, although often only partially (Baskin et al, 1979; Bendayan and Zollinger, 1983; Bendayan, 1984).

1.2   Overview

Those epoxy resins which are of practical value to the microscopist should be cured with heat (60°C or more for at least 16 hours) even after catalytic accelerators have been added. The epoxy resins readily dissolve lipids (especially at 60°C), which coupled with full dehydration and the use of organic solvent intermediaries (clearing agents), necessary for the infiltration of tissue, means that epoxide embedding is extremely extractive (Sects. 1.3.3/2.2.3.2). For this reason tissue is usually post-fixed in osmium which reacts most importantly with unsaturated lipid and thereby helps to preserve membrane structure. Osmium, however, also reacts with many of the other constituents of tissue (e.g. proteins), changing them structurally (Baschong et al, 1984; Emerman and Behrman, 1982) and lessening the ability of antibodies to recognise them for immunolocalisation (Bendayan, 1984; Roth et al, 1981; Hayat, 1986). The necessity for heat polymerisation when using epoxy resins leaves very little room for manoeuvre to lessen the deleterious effects of processing, infiltration and embedding on tissue. Tissue must be very stable before embedding, and, therefore, delicate methods of fixation cannot be recommended.

In general, omitting osmium fixation leaves tissue more reactive histochemically, cytochemically and immunocytochemically. However, organic solvent dehydration and lengthy infiltration with epoxides at room temperature is very extractive, particularly of lipid. Epoxides are infiltrated in to tissue slowly because of their high viscosity, and they can only be cured using heat (Sect. 5.1). In the absence of post-osmication, this can seriously damage tissue ultrastructure. In addition, epoxy resin sections have an unspecific attraction for immunolabels such as colloidal gold leading to high background.

Figure 1. 1% glutaraldehyde perfusion-fixed pituitary of rat embedded in LR White following partial dehydration and the rapid chemical catalytic room temperature method (Sect. 3.3.2). Counterstained with uranyl acetate and lead citrate. The tissue was not post-fixed in osmium. (Mag. x 11,000).

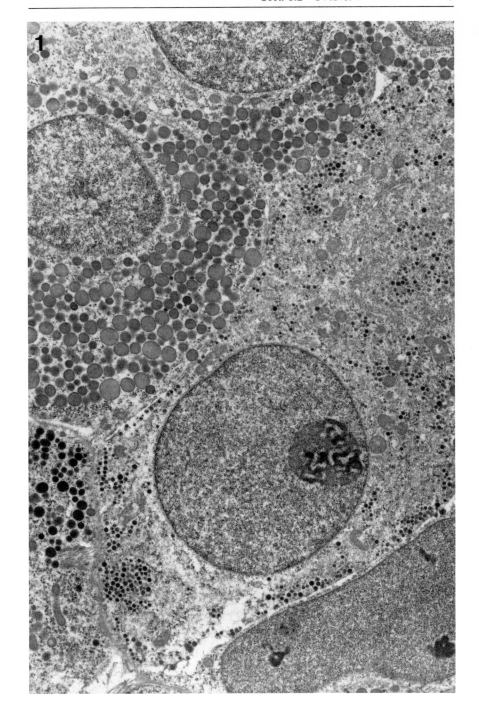

It is for this reason that less deleterious embedding methods using crosslinked acrylic resins, which are stable within the electron beam, have been developed. These techniques enable osmium fixation to be omitted whilst preserving reasonable levels of ultrastructure and good immunolabelling. Simply switching to an acrylic resin such as LR White or Lowicryl offers the advantage of much lower nonspecific backgrounds when immunolabelling sections (Roth et al, 1981), even if the tissue is post-osmicated (Craig and Miller, 1984; Graber and Kreutzberg, 1985).

However, important gains in tissue immunoreactivity can be made if osmium post-fixation is omitted and the primary aldehyde fixation is minimised in concentration and time. Under these conditions, full dehydration and epoxy resin embedding would be much too deleterious. The tissue after fixation with low concentrations of glutaraldehyde (< 0.2%) is now particularly prone to damage by the organic solvents and resins used during the processing steps, and more sensitive embedding procedures are required. In order to minimise extraction, the tissue can be partially dehydrated at room temperature to 70% ethanol before infiltration with, and embedding in to LR White; tissue prepared in this way shows an improved response to antisera and lectins over that shown by fully dehydrated tissue (Newman et al, 1982, 1983a). Further gains with partial dehydration are achieved if the method of polymerisation is changed from heat to chemical catalysis (Yoshimura et al, 1986; Newman et al, 1986; Newman and Hobot, 1987; Newman et al, 1989; Bowdler et al, 1989; Hobot and Newman, 1991) although block size has to be reduced. Low concentration antigenic sites, not detectable after high levels of glutaraldehyde fixation (>1%), are now observed after lowering the fixative concentration to 0.1-0.2% (Hobot and Newman, 1991; Newman and Hobot, 1989). In addition to LR White, Lowicryl K4M can also be employed for the partial dehydration procedure (Hobot, 1990; Hobot and Newman, 1991).

If the antigen is sensitive to fixation or easily extracted, more complicated research methods employing the Lowicryl resins will be necessary. Progressive Lowering of Temperature (PLT) methods achieve embedding in the low viscosity Lowicryl resins K4M and HM20 at -35°C or lower (Kellenberger et al, 1980; Carlemalm et al, 1982a; Armbruster et al, 1982). Chemical reactions are much slower at low temperatures so that when PLT methods are applied to tissue which is only very lightly fixed, there will still be good retention of both ultrastructure and high levels of immunosensitivity (Hobot, 1989). Polymerisation is by ultra-violet (UV) light so that even smaller blocks than are required for LR White are necessary, although the method is unsuitable for pigmented or osmicated tissues (unless chemical polymerisation is employed, Sect. 5.3.2.4).

Cryosubstitution can be used to avoid chemical fixation entirely. The tissue is rapidly frozen and the ice gradually removed by an organic solvent at very low temperature (-80°C). The Lowicryls K4M, HM20, K11M and HM23 remain liquid at -35°C to -80°C and can be infiltrated in to the tissue and polymerised with UV-light while still at these very low temperatures (Acetarin et al, 1986; Carlemalm et al, 1985b; Hobot, 1989, 1991). Freeze-drying also avoids chemical fixation and similarly can benefit from low temperature embedding and polymerisation

(Chiovetti et al, 1986, 1987; Edelmann, 1986; Livesey et al, 1989). Both methods depend on very rapid freezing to produce the vitreous or non-crystalline form of ice necessary to avoid artefacts so that only tiny pieces of unpigmented tissue are suitable.

The adverse affects of organic solvents and resins on retaining antigenicity within tissue can be avoided by using preparative procedures developed by Tokuyasu (1973, 1984) in the field of cryoultramicrotomy. Chemical fixation is still required at the outset, and minimally cross-linked tissue can be prone to extraction during treatment and immunolabelling of the "thawed" cryosections (Hobot and Newman, 1991).

A continuous series of overlapping techniques therefore exists, starting with the least sensitive traditional epoxide method and finishing with highly sophisticated cryosubstitution and freeze-drying methods. These are, therefore, available to provide an appropriate approach to a huge range of biological problems.

1.3 Strategies for Planning a Project

1.3.1 Introduction

The careful planning of a protocol is pivotal to any successful experiment. The strategic approach should encompass not only the tissue processing and resin embedding methods, but also the immunocytochemical method (Part II) to follow.

The flow-chart (Fig. 2) summarises the progression of methods that is available for processing tissue in to resin. Moving across the chart from left to right, the techniques and resins vary as the substance to be localised becomes more difficult to preserve. The stability of the substance to be localised in the presence of chemical fixatives is a very important factor. Chemical fixation inevitably changes the chemical and physical make-up of biological tissue and can, for example, render tissue antigens unrecognisable to their antisera for immunocytochemistry or destroy carbohydrate groups needed for cytochemistry. Various methods are available for testing the robustness of the substance, such as exposing extractions of it on nitro-cellulose to chemical fixatives at different concentrations and times followed by the complete localisation procedure. If the tissue is in the form of a cell suspension, samples of the cells may be adhered to the base of wells (plated-out) in a tissue micro-test plate and different procedures of permeabilisation (if necessary) and fixative concentration and time can again be followed by the preferred localisation procedure. Frequently, important information as to the resistance of cell substances to chemical fixation is acquired, but these analyses are limited to the light microscope and have other draw-backs. Fixatives often react with substances in cells and tissue differently from the way they react with the same substances when they are purified and isolated. Sometimes the substance sought is not easily extractable or may be unknown (as with many monoclonal antisera). If the substance is intracellular, the cell-plating method may not be applicable. Whether such factors apply or not, the most important analysis is the one completed on the processed biological tissue and, of course, no information is provided as to the resistance of the substance to tissue processing methods following fixation.

The demonstration of a substance at a cellular or sub-cellular level starts with tissue, which can be divided, very broadly, in to two categories: (1) free-living cells in to which for example cell cultures, bacteria, algae, protozoa and circulating cells of blood and lymph can be grouped, and (2) solid tissues from clinical biopsies and surgery or from animals. The first category is usually the easier of the two to deal with in that cell suspensions can be formed and, following filtration or centrifugation, samples of the the cells can be resuspended in a variety of fixatives at high or low concentration for various lenghts of time. It has been traditional in microscopy simply to form a pellet of fresh cells which is then fixed and processed as though it were a solid tissue block. However, the substance to be localised must be resistant to change by chemical fixation because higher concentrations of fixative for longer periods of time are required to diffuse through pellets than are needed for cells in suspension. The cells of even lightly chemically-fixed cell-suspensions lose their self-adhesive qualities and cannot be pelleted and processed without considerable loss of cells in the supernatants of dehydrating and washing solutions. This problem can be remedied by resuspending the fixed cells in a matrix such as agar or gelatin and processing once again as though for solid tissue (Sect. 3.1.1). Where it is known that chemical fixation will destroy the substance to be localised, smears of freshly concentrated cells must be rapidly frozen (cryoimmobilisation) and processed through cryotechnology (Sect. 1.3.2.2).

Solid tissue must be diced in to small pieces and immersed in fixative or perfused with fixative either in the whole animal or in an organ (Sect. 3.1.3). Solid tissue blocks, like cell pellets, depend on diffusion to deliver chemical fixatives to their centres. For this reason relatively high concentrations (> 0.5% glutaraldehyde or 2% formaldehyde) of fixative for relatively long fixation times (> 2 h) are usually necessary. The only successful way of using low concentrations of fixative for short times is with perfusion where the vasculature of the animal or organ is used to deliver the chemicals in the most immediate way possible to the tissues, and fixation is no longer dependant on diffusion.

Having chosen a tissue, together with the substance of interest which can be localised within it and the fixative method that has a chance of preserving the reactivity of the substance, it is time to consider the processing method that will provide sectionable material. Generally it is the fixative method which dictates the preferred processing route. For this reason it is important to consider fixation and processing together when planning a project. Where tissue is heavily fixed, for example tightly cross-linked by high concentrations of glutaraldehyde, a method, such as complete dehydration of the tissue in an organic solvent and epoxy, or acrylic resin embedding at room temperature, or using heat for polymerisation, is usually perfectly acceptable, even though it is extractive. Such a method, however, would not be acceptable if the tissue had been only lightly fixed in order to

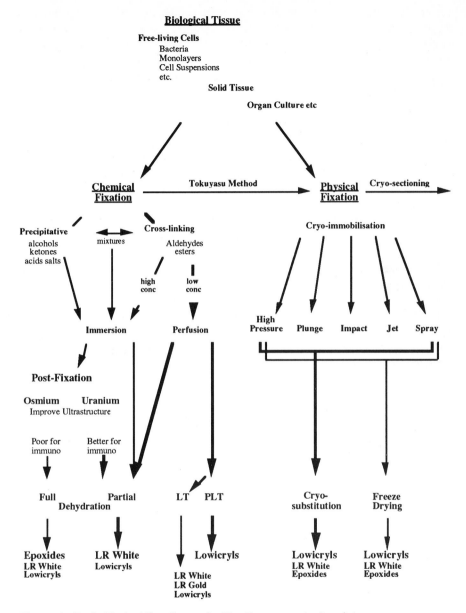

Figure 2. Resin Embedding Strategies For Immunocytochemistry

The relationships between the main technologies available for resin embedding are
shown increasing in complexity and decreasing in tissue block size from left to right.
Preferred routes are indicated by thicker arrows

preserve the reactivity of a specific tissue substance and was, therefore, much more prone to distortion and extraction in organic solvents and resins. To overcome the deleterious effects of organic solvents and resins, the time of exposure to them can be reduced by utilising rapid processing protocols at room temperature. Alternatively, low temperature can provide protection, even with prolonged processing times.

Partial dehydration at room temperature and cold acrylic resin polymerisation could be employed first and, if unsuccessful, lowering of temperature (LT) or progressive lowering of temperature (PLT) methods should follow. Eventually, when all else fails, only cryoimmobilisation and either cryosubstitution or freeze-drying remain as the final resort. It should be remembered, however, that these techniques become increasingly more difficult to perform and the pieces of tissue to which they apply must be made progressively smaller. For this reason, it can be seriously limiting and a considerable waste of time to use a complicated low temperature embedding regimen on heavily fixed tissue.

The applicability of the prefered immunocytochemical method to the resin system should also be considered (Fig. 2). LR White is very versatile and tissue sections will counterstain or immunolabel with almost anything. But, LR White can only be used for embedding down to temperatures of about -20°C. Counterstaining methods using basic dyes or immunoperoxidase/DAB applied to sections of tissue in Lowicryl K4M, however, produce unavoidably high background staining and often uninterpretable results. Lowicryl HM20 will counterstain or immunolabel with immunoperoxidase/DAB very much like LR White, but this resin, unlike Lowicryl K4M, cannot be used for partial dehydration at room temperature with 70% ethanol. All the resins can be used for colloidal gold/silver amplification techniques.

Bearing in mind that the regime which is finally chosen should be considered as a whole i.e. tissue type/fixation method/dehydration method/resin and resin polymerisation method, it is possible to dissect the various options and examine them separately.

1.3.2 Fixation Strategies

Tissue fixation can follow one of two paths: the tissue can be stabilised by chemical cross-linking and/or precipitation, involving the use of chemical fixatives; or it can be physically immobilised in ice by using cryotechniques. Both procedures allow for subsequent resin embedding of the tissue.

1.3.2.1 Chemical Fixation

Introduction
The choice of processing technique and resin is intimately connected with the method of tissue fixation. Fixation and resin processing, therefore, have to be

considered as a whole. Generally, in glutaraldehyde fixed tissue, antigen sensitivity is inversely proportional to the concentration of the glutaraldehyde and the time for which it is used (Hayat, 1986). Fixation in high concentrations of glutaraldehyde (>1%) for periods of time exceeding 15 min have been shown to be deleterious to antigenic response (Hobot and Newman, 1991). The loss is irretrievable and there is nothing to be gained from using sophisticated resin embedding methods. Lowering the glutaraldehyde concentration (to 0.1-0.2%) and processing at room temperature with full dehydration of tissue also does not improve the tissue's antigenic response. The tissue is now minimally stabilised and more prone to extraction and conformational damage from the effects of room temperature full dehydration and heat polymerisation. Minimising fixation in glutaraldehyde in order to reveal sensitive, low concentration or secondary, antigenic sites, therefore, necessitates the use of more complex and less deleterious dehydration methods, such as partial dehydation at room temperature or PLT, to maintain the tissue's improved antigenic reactivity (Hobot and Newman, 1991; Newman and Hobot, 1989). These methods are coupled to the use of acrylic resins and to the possibilities afforded by them in avoiding heat polymerisation completely (Tissue Fixation, see below; Chap. 5).

Tissue Fixation
With the exception of cryosubstitution and freeze-drying (molecular distillation drying) methods, tissues must be chemically stabilised (fixed) before they are processed in to resin. The extent of fixation can be varied and depends on the harshness of the conditions of dehydration, infiltration and polymerisation employed. When employing a protein crosslinker, such as glutaraldehyde, the end-point of fixation, which is the irreversible crosslinking of all the tissue protein, is probably its most stable condition and, therefore, desirable for structure. However, "mild fixation" is generally preferred since the level of antigenicity preserved is often inversely proportional to the concentration of the fixative and the duration of its use (Hayat, 1986), but may result in poor structural preservation. In practice, a compromise has to be struck between total chemical stabilisation of tissue, good for ultrastructure, and its opposite, total avoidance of chemical change which is good for antigenicity. The definition of "mild fixation" is very nebulous, being any fixation regime which stops short of complete chemical stabilisation of tissue. Mixtures of glutaraldehyde/formaldehyde (e.g. 0.1%/2%) have been advocated on the basis that formaldehyde penetrates faster and does not crosslink protein like glutaraldehyde. Such mixtures have been found to give the same level of antigen preservation within tissue as 1% glutaraldehyde (Bowdler, 1991).

Neutrally buffered glutaraldehyde (2%) diffuses through tissue very slowly indeed (approximately 0.5 mm after 3 hours; Hayat, 1981) so that only small blocks (< 1 mm^3) are suitable. Autolysis may damage the centre of larger blocks. Formaldehyde penetrates tissue faster but is slow to crosslink protein. When neutrally buffered, it forms only reversible peptide bonds for about the first 20 hours of its application. Prolonged buffer washes (overnight) can lead to reversal of fixation with extraction and loss of structure. With fixation times of over 20 hours, irreversible methylene bridges are eventually formed. The mixtures of glutaraldehyde with larger concentrations of formaldehyde (for example neutral buffered 0.1% glutaraldehyde with 2% formaldehyde mentioned above) have been popular in the hope that the mixture would exhibit the best properties of its components. The components, however, can vary widely in their composition. Commercial forms of glutaraldehyde and formaldehyde contain cyclical polymers and numerous impurities (Gillett and Gull, 1972). Immersing tissue in the more rapidly penetrating mixtures makes it very difficult to control the extent of tissue cross-linking, and experimental reproducibility is impossible. In addition, the results of one laboratory cannot be compared with those from another. More consistent results can be obtained by employing high purity reagents. Highly purified, monomeric glutaraldehyde should be stored, as concentrated as possible (it can be purchased at 25%, 50% or 70%; Polysciences) at a temperature of at least -20°C because it polymerises rapidly at room temperature, particularly when diluted, with the evolution of glutaric acid. On the other hand, high concentrations of purified formaldehyde, which can be purchased or prepared from paraformaldehyde, must be diluted and left for long periods before they depolymerise.

Optimising tissue reactivity is, therefore, crucially dependant on minimising the effects of fixation, i.e. lowering the glutaraldehyde concentration to 0.1-0.2%. Kraehenbuhl et al (1977) observed *in vitro* that 1% glutaraldehyde lowers the reactivity of antigenic sites more than concentrations of less than 0.2%. With 1% solutions there is probably a stronger cross-linking of proteins or enzymes, so reducing their immunoreactivity. Using glutaraldehyde solutions of 0.1-0.2% for fixation puts ultrastructural integrity at risk when the tissue is subsequently dehydrated in an organic solvent. Such delicately fixed tissue is therefore best processed by one of two dehydration methods, known to be less deleterious in causing extraction of soluble tissue components and to changing protein conformational structures (Carlemalm et al, 1982a; Newman and Hobot, 1989; Hobot and Newman, 1991): (i) partial dehydration up to 70% organic solvent at room temperature, or, (ii) progressive lowering of temperature techniques (PLT) at -35°C to -50°C (see Chap. 2).

The method of resin polymerisation will also affect the retention of immunoreactivity within the tissue, and must be chosen accordingly (Sect. 1.3.4). For example, LR White and the Lowicryl resins can be polymerised with heat at 50°C, but, when embedded by this method, delicately fixed tissue will be damaged during processing. LR White and the Lowicryls can also be catalytically polymerised and less stable tissue than would be advisable for heat polymerisation can be successfully processed. The Lowicryls, however, can be polymerised with UV-light at temperatures far lower than would be applicable with LR White or LR Gold, (both can be UV-light polymerised but only down to about -25°C after which they freeze), so that methods restricting tissue stabilisation to even lower levels can be employed. In fact PLT methods can only preserve the ultrastructural integrity and immunoreactivity of tissue, at a level significantly above that provided by far simpler methods, if it is fixed in the lightest way possible.

Modes of Fixation

Cell suspensions, bacterial and cell cultures and monolayers are relatively easy to handle by choosing the appropriate method (see Sects. 3.1.1/3.1.2 for practical details; Hobot, 1991; Hobot and Newman, 1990). They can be fixed briefly in low concentrations of aldehyde, for example 0.1% neutral buffered glutaraldehyde for 10 minutes. If highly purified reagents are used, experiments at different concentrations of fixative for different times can be reproducibly repeated (Hobot and Newman, 1991). The problem of transferring the cell suspensions from one solution to the next is easily solved. The cell suspension can be harvested or concentrated by centrifugation or filtration and the cells resuspended in agar or in a low melting point agarose or gelatin mixture (Hobot et al, 1984, 1985; Kellenberger et al, 1972). After cooling, the solid matrix can be cut in to small blocks, which can be processed further as if they were actual tissue pieces. Fixation can be performed before this treatment (preferable) or after (Sect. 3.1.1).

Dealing with solid tissue is much more difficult (Sect. 3.1.3). Fine control of fixation is needed for delicate, fixative-sensitive antigens or for demonstrating low-concentration sites of more robust antigens (Hobot and Newman, 1991). It is difficult to restrict the extent of fixation when employing immersion methods as immersion-fixation does not lead to the even distribution of the fixative throughout the tissue. The outside of tissue blocks is inevitably exposed to the fixative for far longer than the centre because of the time involved in diffusion. There is a gradient from the outside of the tissue block towards the inside of more cross-linked to less cross-linked tissue. The gradient for antigenic reactivity of such immersion-fixed tissue is often in the reverse direction. This can lead to an eratic localisation of antigenic sites within tissue. Perfusion-fixation can overcome these problems, especially when low concentrations of fixative are employed (see below). However, for immersion-fixation if the size of the tissue samples is kept small in at least one dimension, and the fixative used is based on neutral-buffered formalin, it is possible to achieve reasonable levels of fixation. For example, 2% formaldehyde (5% formalin, methanol free) to which has been added 0.1% purified glutaraldehyde (Polysciences) when used for short times (< 3 hours, but to be determined empirically) has proven of some value for preserving tissue reactivity, especially when followed by post-fixation in uranyl acetate, but leaves the tissue vulnerable to extraction and polymerisation distortion. In addition, prolonged buffer washes must be avoided or fixation reversal will result. But, for the preservation of delicate or low concentration antigenic sites there is a need for very low concentrations of glutaraldehyde. However, when used for immersion-fixation methods, the uniformity and quality of structural preservation achieved is often poor irrespective of the processing and resin methods chosen.

Precise, uniform and reproducible levels of fixation can be achieved with vascular perfusion which delivers the fixative to large volumes of tissue almost instantly. Perfusion removes the dependence that immersion fixation has on diffusion so that low concentrations of fixatives such as glutaraldehyde (0.1-0.5%) can be used for short periods of time. With vascular perfusion it is possible to observe the effects of varying the concentration, temperature and time of fixation on the preservation of ultrastructure and tissue reactivity after different processing and embedding procedures (Newman and Hobot, 1989; Hobot and Newman, 1991).

If glutaraldehyde is used at 1% or above for even very short periods of time (15 min) at room temperature (22°C) heavy crosslinking of protein occurs and the need for delicate embedding measures is removed (Newman and Hobot, 1989; Hobot and Newman, 1991). Only if the concentration of glutaraldehyde is controlled at between 0.1-0.4% for 15-120 min at 22°C, leaving the tissue vulnerable to the subsequent processing and polymerisation steps, are more delicate methods required. For the preservation of structure and antigen reactivity in tissue fixed in such low concentrations of glutaraldehyde, partial dehydration or low temperature (PLT) methods will probably be necessary. To effect fine control of fixation pure reagents must be used. They will not necessarily improve the quality of fixation, but they guarantee a high degree of experimental reproducibility and make the results obtained by one laboratory comparable with those of another (Newman, 1987).

During perfusion-fixation, the fixative is delivered to individual organs or the whole animal via the vascular system. In the perfusion of whole small animals, after deep anaesthesia, the heart is catheterised and hydrostatic pressure drives the fixative around the body. For larger animals a peristaltic pump is necessary. Low concentrations of fixative are easy to administer and the rapid and intimate distribution of it throughout the tissues ensures that problems of penetration and the creation of the kind of diffusion gradients seen during immersion fixation do not occur. The length of time over which fixation is conducted can be varied down to the practical limit of about 10 minutes.

Antigenic sites within tissue that still cannot be detected following these very sensitive perfusion-fixation procedures with low concentrations of glutaraldehyde, will necessitate the use of more complex cryotechniques preferably in combination with the Lowicryl resins.

1.3.2.2 Cryoprocedures

Introduction
Fixatives affect tissue in two ways: (i) by chemical cross-linking of cellular components, which can lower or remove antigenic reactivity, and (ii) by altering membrane permeability and inducing ionic leakage (Woldringh, 1973). Glutaraldehyde induces a one-way leakage of ions out of the cell, whilst osmium tetroxide causes a breakdown in membrane permeabilty such that the internal ionic concentration or osmotic pressure equilibrates with that of the external milieu (Woldringh, 1973). These changes can lead to rearrangements in cellular structure and organisation (Hobot et al, 1985). Rapid freezing techniques (grouped under the heading "cryofixation"; for reviews see Robards and Sleytr, 1985; Steinbrecht and Zierold, 1987) avoid chemical pre-fixation, giving a different perspective on ultrastructure and offering the opportunity to preserve in tissue both structures and

antigenic sites which are sensitive to even low concentrations of fixative. In fact, tissue is not fixed in the way that chemical cross-linking is achieved by using aldehydes or osmium, but rather physically immoblised in ice, i.e. *cryoimmobilisation*. The term "cryofixation" may therefore be a misnomer in the context of its present use in the literature and should be replaced by *cryoimmobilisation* (cf. Hermann and Müller, 1991).

The advantage of cryotechniques to freeze biological tissue rapidly, and via the various processing procedures available, to preserve faithfully the specimens' structure and antigenic reactivity, is offset by the size of the samples that can be frozen successfully. Tissue block sizes have to be much smaller and thinner than those required for chemical fixation (0.1-0.2 mm vs. > 1 mm^3; Sects. 1.3.1/4.1).

Cryoimmobilisation and Resin Embedding

The frozen tissue can be processed in to resin through two routes:

(a) Cryosubstitution replaces the ice in the tissue, at very low temperature, with an organic solvent which is miscible with the embedding medium. Infiltration and embedding can then take place at any temperature between 60°C and -80°C.

(b) Freeze-drying sublimes away the ice leaving completely dry tissue in to which a resin can be infiltrated, with embedding again at any temperature from 60°C to -80°C.

In order to avoid changes to the tissue induced by returning it to room temperature or higher, the Lowicryl resins provide a means of embedding at low or very low temperature. Lowicryl K4M can be used down to -35°C, Lowicryl HM20 down to -50°C, Lowicryl K11M down to -60°C and Lowicryl HM23 down to -80°C. In fact as the temperature of organic solvent-substituted or freeze-dried tissue rises so does the "activation energy", leading to the aggregation, precipitation or collapse of biological macromolecules. This appears to be accelerated in the temperature range -58°C to -10°C (MacKenzie, 1972) so that Lowicryls K11M and HM23 may be particularly important for low temperature embedding of cryoimmobilised tissue (Acetarin et al, 1986; Acetarin and Carlemalm, 1985; Carlemalm et al, 1985b; Hobot, 1989). Impregnation of tissue with resin may be difficult at these low temperatures so that protocols with extensive infiltration times may have to be formulated.

Cryosubstitution is advantageous in dealing with fixation/dehydration sensitive structures, as for example prokaryotic DNA, whose structure cannot be stabilised by chemical cross-linking so rendering it sensitive to collapse and aggregation in organic solvents. Further, its structural integrity is dependent upon maintaining the ionic balance within the cell, disrupted during chemical fixation (Kellenberger, 1991; Kellenberger and Ryter, 1964). Results have indicated that the level of structural preservation that can be achieved by cryosubstitution is not only much superior to that obtained with conventional or PLT techniques (Hobot et al, 1985), but is the same irrespective of the substitution medium or embedding resin used (Epon at 60°C, [Hobot et al, 1985]; Lowicryl K4M at -35°C, [Hobot, 1990; Hobot et al, 1987]; Lowicryl HM20 at -40°C, [Dürrenberger et al, 1988]; Lowicryl K11M at -60°C, [Villiger, 1991]; Lowicryl HM23 at -70°C/-80°C, [Hobot, 1989; Villiger, 1991]). No published information is currently at hand for the effects of freeze-drying (using conventional freeze-drying apparatus or the "Lifecell" Molecular Distillation Dryer) on the preservation of prokaryotic DNA or other similarly sensitive structures. However, initial results for bacteria with the "Lifecell" Molecular Distillation Dryer are encouraging (J.A. Hobot, unpublished results).

The important point to bear in mind with these results, and the discussion of the possibilities of collapse or aggregation of tissue components occurring during the substitution or drying cycle, especially in the temperature range of -58°C to -10°C (MacKenzie, 1972), is that biological systems do not behave as individual test solutions, but as complex interactions of solutes, ions, proteins etc. Thus, different biological specimens will behave differently during the time the temperature is raised, and can, consequently, have varying levels of cellular preservation. From results presented in the literature and of published discussions on the influences of the various processing steps (Artymink et al, 1979; Carlemalm et el, 1982a; Edelmann, 1986, 1989a, 1989b, 1991; Frauenfelder et al, 1979; Humbel et al, 1983; Kellenberger, 1991; MacKenzie, 1972; Schwarz and Humbel, 1989), it would probably be best to maintain the specimens temperature below -30°C. Bacterial DNA is a most sensitive indicator of what can happen if the appropriate "protection" is not given to the specimen during the processing steps of either cryosubstitution or freeze-drying. The "protection" can be for example the presence of a fixative in the substitution medium for temperatures above -30°C. Bringing the specimen to temperatures above -10°C, usually up to room temperature, can be deleterious for the tissue. Indeed, it would be of greater merit to maintain the advantages offered by low temperatures, avoid the use of fixatives, and proceed with the processing, embedding and polymerisation of the tissue below -30°C.

Cryosubstitution
When combined with final embedding in epoxides, cryosubstitution has led to some small but significant advances in our understanding of ultrastructure (fungi - Hoch and Howard, 1980; Howard, 1981; Dahmen and Hobot, 1986; Hippe and Hermanns, 1986; bacteria - Ebersold et al, 1981; Amako et al, 1983, 1988; Hobot et al, 1984, 1985; cartilage matrix - Hunziker and Herrmann, 1987; epithelial cells - Sandoz et al, 1985; see Hobot, 1989, 1990, 1991 and Nicolas et al, 1991 for a more detailed discussion). These "conventional" forms of cryosubstitution included fixatives (osmium, glutataraldehyde, uranyl acetate) in the substitution solvent. Any osmium in the substitution medium will result in the necessity of its removal from the tissue by pretreatment of the resin sections (Sect. 8.1.5), and could also lower the gains made in retaining immunoreactivity within the tissue which was one of the major goals in initially employing cryotechniques. It is perfectly possible to omit all fixatives, however, when cryosubstitution is combined with low temperature Lowicryl embedding (Humbel et al, 1983; Humbel and Müller, 1984; Hobot, 1989, 1990; Hobot et al, 1987; Edelmann, 1989a, 1989b; Monaghan and Robertson, 1990; Schwarz and Humbel, 1989). Based upon the findings of MacKenzie (1972)

and results from Hobot (1989, 1990) discussed earlier (preceding section), the addition of fixatives to the substitution medium can be avoided when the tissue is embedded below -30°C.

The advantages of cryosubstitution for an improved immunocytochemical response over PLT have been demonstrated (Carlemalm et al, 1985b; Schwarz and Humbel, 1989), but in cases where partial dehydration or PLT were preceded by a very sensitive approach using low concentrations of glutaraldehyde in combination with perfusion-fixation, the same levels of antigenic reactivity were observed (Hobot, 1989; Hobot and Newman, 1991; Newman and Hobot, 1989). The improved structural preservation achieved by cryosubstitution has led, for example, to a more accurate immunolocalisation of bacterial DNA and its related proteins (Björnsti et al, 1986; Dürrenberger et al, 1988; Hobot et al, 1987); of proteoglycans in the extracellular matrix of rat cartilage (Hunziker and Herrmann, 1987); and of botanical specimens (Kandasamy et al, 1991). Where fixation-sensitive antigens have been alleged to have been detected by cryosubsititution (Monaghan and Robertson, 1990), other possibilities following chemical prefixation and resin embedding were not exploited. Some antigens, such as membrane-bound cell surface receptors, are genuinely sensitive to low concentrations of aldehydes. If these are not detectable after PLT embedding in Lowicryl HM20, which can preserve lipids very well (Weibull and Christiansson, 1986; Sects. 1.3.3/2.2.3.2), the exploitation of cryotechniques and cryosubstitution will be necessary.

Freeze-Drying
The technique of freeze-drying has been applied to tissue that was subsequently osmicated (by osmium vapour) and embedded in Epon/Araldite (Dudek et al, 1981, 1982, 1984) or Spurr's resin (Linner et al, 1986). However, using Lowicryl K4M as an embedding medium, it does not show a clear advantage for immunocytochemistry when compared with PLT (Chiovetti et al, 1987; Jorgensen and McGuffee, 1987). Freeze-drying followed by low temperature embedding in the Lowicryl resins (Lowicryl HM20 - Wroblewski and Wroblewski, 1986; Lowicryl HM23 - Wroblewski et al, 1990; Lowicryl K11M - Edelmann, 1986) has been used in the field of electron microscope X-ray dispersive microanalysis. But here it would seem that cryosubstitution provides a firmer basis for analytical techniques by stabilising better the cellular ionic environment (discussed in Edelmann, 1989a, 1991).

Freeze-drying avoids the use of organic solvents, although collapse or aggregation of tissue substances can occur during the drying procedure (MacKenzie, 1972), with soluble cytoplasmic components adhering to membranous cellular structures (discussed above - Sect. 1.3.2.2 for Cryoimmobilisation and Resin Embedding). This is probably more significant if the dry tissue is brought up to room temperature and subjected to vapour fixation. A possible drawback when using vapour fixation is that the vapour may inadvertently introduce water vapour, its source coming from the fixative (e.g. osmium crystals). This could cause a certain level of tissue rehydration to occur. Another possibility is the redistribution of ions, due to the release of bound water from the tissue by the interaction between protein and osmium (Edelmann, 1986). Finally, the biological tissue is still subjected to the extractive and disruptive influence of an organic resin at room temperature. At -30°C or below, embedding is probably less deleterious, but the high membrane lipid retention following freeze-drying can create a serious barrier to resin infiltration.

A new and similar process, molecular distillation drying, may overcome some of the objections to freeze-drying (Linner et al, 1986; Linner and Livesey, 1988). Freeze-drying by this method is started below the usual drying temperatures of -80°C to -90°C. Removal of water from biological specimens at temperatures of -125°C to -100°C seems to be possible (Livesey et al, 1991). Complex control of temperature and high vacuum is required and this is really only possible with dedicated equipment ("LifeCell Process", RMC, Tuscon, Arizona, USA). Early results have been with osmicated tissue embedded in Spurr's resin (Linner et al 1986; Livesey et al, 1989; Jesaitis et al, 1990; VanWinkle, 1991) but little has been done with the low temperature acrylic resins, except in the field of biological X-ray microanalysis (discussed in Edelmann, 1986). It is in this context that the above-mentioned initial results for bacteria with the 'Lifecell' Molecular Distillation Dryer are encouraging (J.A. Hobot, unpublished results). However, to evaluate its potential in immunomicrocopy, an in-depth comparative study of freeze-drying, cryosubstitution and the chemical fixation methods detailed in this book are required. The present use of osmium vapour and epoxy resins should be avoided (Edelmann, 1986; Sects. 1.2/4.6), as also should the use of the temperature range above -10°C where collapse/aggregation phenomena are most likely to occur (MacKenzie, 1972), for these are rather retrograde steps to take with such an exciting new low temperature development.

Cryoultramicrotomy for Immunocytochemistry
The technique was first introduced by Tokuyasu (1973), and relies upon pre-fixed tissue being infiltrated with sucrose, frozen, and sectioned at low temperatures (e.g. -120°C to -80°). Afterwards, the sections are allowed to thaw out, and the "thawed" tissue sections/slices are subsequently immunolabelled (Tokuyasu, 1984). Cryoultramicrotomy is a rapid technique that yields blocks for sectioning very quickly (e.g. after approximately 30-60 min following fixation, the time depending upon how long sucrose infiltration lasts). This should be contrasted with rapid embedding procedures in to acrylic resins (Sect. 3.3.2), which, following fixation and short buffer washes, can only produce blocks ready for sectioning after 2.5 hours, but it is easier and more convenient to store resin blocks than it is to keep frozen blocks under liquid nitrogen.

Cryoultramicrotomy for immunocytochemistry is a methodology that does not require organic solvent dehydration or resin embedding of tissue. It has, therefore, been suggested that as these two deleterious steps are avoided, cryoultramicrotomy is the most advantageous technique to employ. Chemical fixation, however, with all of its inherent problems (Sect. 1.3.2), is still required. Generally, in order to retain structural integrity, (i.e. to prevent the frozen tissue sections from falling apart when thawed for subsequent passage through buffer/immunoreagent droplets), relatively high levels of fixative concentration have to be used. Low concentrations of glutaraldehyde (0.2%) allow the immunolabelling of low concentration antigenic sites in tissue prepared by partial dehydration or PLT protocols followed by resin embedding, but often such sites do not show immunolabelling following cryoultramicrotomy (Hobot and Newman, 1991). Similar inferences can be drawn from the failure to detect low concentration sites of amylase in rat pancreas by cryoultramicrotomy using mixtures of 0.2% or 0.5% glutaraldehyde/2% formaldehyde (Posthuma et al, 1984, 1986; Slot and Geuze, 1983). Structural proteins are probably preserved, at least to some extent, by low concentrations of fixative, but, because the frozen sections must be thawed and incubated on buffer/reagent droplets, there may well be extraction of smaller peptides and more soluble proteins. Increasing the extent of cross linking within tissue, by raising the fixative concentration, will reduce this loss and improve the structural preservation of the tissue, but only at the expense of antigenic sensitivity.

The method of fixing the tissue, whether by perfusion or immersion techniques, may be important to the final outcome of observing the location of antigenic sites. For low concentrations of fixative (< 0.2% glutaraldehyde), perfusion methods are more desirable. Frozen tissue sectioned after 0.2% glutaraldehyde perfusion-fixation, thawed, and immunolabelled, showed better structure than immersion-fixed tissue. Furthermore, the perfusion-fixed tissue sections did not fall apart upon thawing as easily as tissue sections of immersion-fixed specimens (J.A. Hobot, unpublished results). Using higher fixative concentrations for immersion-fixation leads to the associated problems of uneven fixation throughout the tissue and the subsequent irregular localisation of the more sensitive antigenic sites (see Modes of Fixation under Sect. 1.3.2.1,).

The ultrastructure that results from the Tokuyasu method is often disappointing, especially when compared to data from cryosubstitution or PLT. The advantage with the method is that, generally, the sites that can be immunolabelled show an improved immunolabelling response when compared with those resin embedding procedures which utilise pre-fixation of tissue with high concentrations of fixative (> 1% glutaraldehyde). As a result, improvements to this technique are being actively sought to raise the level of ultrastructural preservation, and at the same time to detect low concentration antigenic sites within tissue by using low concentrations of fixative or an appropriate combination of fixatives in mixtures.

1.3.3 Dehydration Strategies

Water is the main component of all biological tissue. Most epoxy resins used in microscopy are not miscible with water. The water is removed from the tissue by a dehydrating agent (organic solvent) at room temperature in an ascending gradient of concentration to avoid osmotic shock. Acetone (if stood over Molecular Sieve, Type 4A, 1/16" pellets; Merck) can be used to dehydrate small pieces of tissue and is miscible with epoxy resins. Alcohols, in particular ethanol, are also popular as dehydrating agents but sometimes an intermediary, to assist resin infiltration, is necessary. If the intermediary can be bound in to the resin during polymerisation it is called a reactive intermediary and does not need such thorough removal from the tissue as a non-reactive intermediary. Propylene oxide is a reactive intermediary for epoxy resin embedding. However, if too much of it is left in the tissue, it will change the characteristics of the resin, often leaving it too soft to section easily. Xylene, chloroform and 1,1,1 Tri-chloroethane ("Inhibisol", Penetone; Agar Scientific) are examples of non-reactive intermediaries.

Water miscible extracts of epoxy resin have been advocated as water-soluble embedding media not requiring organic solvent dehydration. Durcupan (Agar Scientific; Fisons Scientific Equipment; Polysciences; Taab) remains the sole commercial example of these and even after modifications made by Kushida (1964, 1966) produces a rubbery final block with poor sectioning characteristics. It can, however, be used as a reactive dehydration intermediary, replacing water before the tissue is embedded in another epoxy resin (Kushida, 1963; Stäubli, 1963; Glauert, 1975). Dehydration in Durcupan may well prevent the loss of some tissue components that are extracted by other organic solvent dehydration techniques. Unfortunately there is evidence that it acts as a fixative of proteins and nucleic acids (Gibbons, 1959; Leduc and Bernhard, 1962; Glauert, 1975) and is, therefore, of little value for immunocytochemistry.

Acrylic resins are often more tolerant of water than either epoxy or polyester resins. Some methacrylates are, like Durcupan, water miscible and do not require organic solvent dehydration. Glycol methacrylate (GMA, HEMA) is hydroxyethyl methacrylate and one of the earliest to be described (Rosenberg et al, 1960). Tissue is dehydrated in a gradient of increasing GMA concentration. A number of proprietary embedding kits based on GMA are available (Agar Scientific; Fisons Scientific Equipment; Polysciences; Taab) but sections cut from GMA blocks are not electron beam resistant, lose mass rapidly, and break in the electron microscope. Those acrylics which do not have some electron beam stability will not be considered here.

The melamines represent a third group of modern water-miscible resins (Shinagawa, 1972; Shinagawa and Shinagawa, 1978) that like Durcupan have been advocated as a means of avoiding the extraction caused by organic solvent dehydration. Melamine resins have been produced commercially as Nanoplast (Frösch and Westphal, 1985; Westphal et al, 1985) and ultrastructurally they have provided some remarkable results (Bachhuber and Frösch, 1983; Frösch and Westphal, 1989; Shinagawa and Shinagawa, 1978). However, disappointingly, tissues embedded in proprietary brands of melamine such as Nanoplast or Nanostrat (Agar Scientific; Fisons Scientific Equipment; Polysciences; Taab) have shown little or no immunoactivity and, as the resins cannot be removed, have no application in post-embedding immunocytochemistry (Frösch and Westphal, 1989).

LR White, LR Gold and Lowicryl K4M (Chap. 2) are not fully miscible with water but can tolerate up to 10-12% water by weight. This characteristic can be used to advantage to avoid full dehydration, which is very extractive. Partial dehydration was first suggested by Idelman (1964, 1965) for embedding osmicated tissue in to the epoxide Epon 812 which is compatible with 70% ethanol in a 1:1 ratio. Unfortunately, although it was claimed that lipid extraction was reduced, little structural difference was seen. It is unlikely that this method would improve immunocytochemistry because of the fixative effects of epoxy resin when in the presence of water (Gibbons, 1959). In any event, no information is available as to the immunoreactivity of tissue prepared in this way.

For acrylic resins, partial dehydration at room temperature proceeds as far as 70% organic solvent. Before infiltration with pure resin, a 2:1 mixture of acrylic resin:70% solvent is employed to lessen osmotic shock. Here the concentration of water is 10%. The most polar acrylic resins probably complete the dehydration of tissue during infiltration (LR White, Lowicryls K4M and K11M). If performed rapidly (1-1.5 hours), high levels of immunoreactivity within the tissue can be retained (Sect. 1.3.1). When low levels of fixation are used (0.2-0.5% glutaraldehyde for 15-60 min), partial dehydration of tissue, by minimising extraction of tissue components, leads to labelling of low-concentration or fixative-sensitive antigenic sites (Sect. 1.3.2.1, Tissue fixation; Newman and Hobot, 1989).

Similar or greater levels of antigenic reactivity can be retained within tissue when employing full dehydration, but only by carrying out the steps at low temperatures (-35°C) with the PLT technique (Newman and Hobot, 1989). At -35°C organic solvents are less deleterious to protein structure and activity (Sect. 2.2.3.2), and the Lowicryl class of acrylic resins were developed to take advantage of this fact.

Summarising very briefly, immunolabelling of fixation-sensitive antigenic sites or low concentration antigenic sites preserved by low concentrations of glutaraldehyde are better detected after either partial dehydration and embedding in LR White or Lowicryl K4M, or after PLT techniques and embedding in the Lowicryls (Hobot and Newman, 1991). New aspects of ultrastructure have also been demonstrated by these methodologies (Hobot, 1990; Hobot et al, 1984; Newman and Hobot, 1987; Roth et al, 1985; see Chap. 2).

A possible explanation for the similar results obtained from partial dehydration and PLT may be that the value of the dielectric constant for 70% organic solvents at 20°C (e.g. methanol 46.3) is close to the value for 100% organic solvents at -35°C (e.g. methanol 46.9; tables and values to be found in Douzou et al, 1976). At -35°C it has been postulated that the lowered dielectric constant during dehydration creates an environment for preserving hydration shells around protein structure (Kellenberger et al, 1980). The similarity of dielectric constant between 100% organic solvent at -35°C and 70% organic solvent at 20°C suggests that hydration shells can also exist during partial dehydration at room temperature (Hobot and Newman, 1991). This would protect proteins, and possibly other tissue components, from the deleterious effects of the organic dehydrating solvent, lessening factors such as extraction, conformational change, aggregation or collapse (precipitation), which would make the tissue antigen less recognisable to its specific antibody.

Such an argument is supported by similar results for retaining levels of immunoreactivity in unfixed biological tissue that had been rapidly frozen and processed by cryosubstitution in to an acrylic resin via an organic solvent at low temperatures (-80°C to -35°C, dielectric constant values for methanol ranging from 62.0 to 46.9; Hobot and Newman, 1991). Rapid freezing and cryosubstitution avoid many of the problems associated with chemical fixation of tissue (Sect. 1.3.2.2). In the case of cellular structures sensitive to the presence of organic solvents, which can cause their precipitation or aggregation, cryosubstitution preserves their ultrastructural integrity (Sect. 1.3.2.2, Cryoimmobilisation and Resin Embedding; Hobot et al, 1985).

Organic solvents extract lipid from tissue fixed with glutaraldehyde but do so less at low temperatures, especially if used up to concentrations of 90%, rather than 100%, at -35°C with the PLT technique (in 90% ethanol, there is 27% extraction of lipids from *Acholeplasma laidlawii* cells; in 100% ethanol, 62% extraction; results from Weibull et al, 1983). Employing PLT down to temperatures of -50°C reduces the amount extracted to 4.2% for ethanol, 12.2% for acetone, and 17.5% for methanol (Weibull and Christiansson, 1986). The choice of resin will influence the final result, with Lowicryl K4M being more lipid extractive than Lowicryl HM20 (Weibull et al, 1983; see also Sect. 2.2.3.2 for an extension of this discussion to embrace the effects of resin on lipid retention). Further, evidence from the work of Zini et al (1989) on the localisation of nuclear phospholipids, minor components of chromatin, suggests that lipid extraction is low after partial dehydration in ethanol and subsequent embedding in LR White.

1.3.3.1 Choice of Dehydrating Solvent

Dehydrating solvents used in microscopy differ in their properties. Generally ethanol is preferred for partial dehydration or PLT, and acetone for cryosubstitution but for optimum results, the dehydrating solvent may have to be matched in its (non-)polar properties to the embedding acrylic. The following points should be borne in mind when deciding upon which organic solvent to use:-

(i)   With acetone, there is no need for an intermediate organic solvent step for successful epoxy resin infiltration.

(ii)   Any remaining acetone can adversely affect acrylic resin polymerisation (Chap. 5). Replacement of acetone during the resin infiltration steps must be complete.

(iii) As 70% acetone has a freezing point of -27°C, it cannot be used in PLT for the 70% organic solvent step at -35°C. For its application in PLT see Weibull and Christiansson (1986).

(iv)   For PLT down to -35°C, there is less lipid extraction if dehydration finishes at 90% ethanol (Weibull et al, 1983).

(v)   For PLT down to -50°C and Lowicryl HM20, both acetone and ethanol are less extractive of some membrane lipids than methanol. 1:1 mixtures of Lowicryl HM20 and 100% methanol extract up to 23% of lipid out of the total (45.6%) that is lost during dehydration and infiltration (Weibull and Christiansson, 1986).

(vi) For cryosubstitution media, acetone is less extractive of membrane lipid than methanol (Weibull et al, 1984). Humbel et al (1983) have reported that methanol is more effective than acetone in the removal of ice at the very low temperatures required for cryosubstitution (-80°C). They found that methanol can tolerate the presence of more water than acetone (10% vs. 1%), and still remain an effective substitution medium. The addition of molecular sieve to acetone, however, improves its performance (Sect. 4.3.1).

(vii) Ethylene glycol and dimethylformamide (DMF) are not miscible with Lowicryls HM20 and HM23. Ethylene glycol and 70% DMF are not miscible with LR White.

(viii) Ethylene glycol can be used for PLT, but only up to a 90% concentration because 100% ethylene glycol has a freezing point of only -12°C.

(ix) 70% organic solvents (used for partial dehydration) are only miscible in a ratio of 1:2 (solvent/resin) with the more polar acrylic resins (LR White, LR Gold, Lowicryls K4M and K11M). Lowicryls HM20 and HM23 are miscible in a ratio of 1:1 with 90% organic solvents.

(x)   With regard to PLT and cryosubstitution, Lowicryl HM20 is less lipid extractive than Lowicryl K4M (Weibull et al, 1983; Sect. 2.2.3.2).

1.3.4 Polymerisation Strategies

In practice there are four ways to polymerise resins over a temperature range of 60°C to -80°C: (i) by heat, (ii) by chemical catalysts, (iii) by ultraviolet light and (iv) by blue light. A fifth way using an electron source is largely impracticable. Heat is the optimal way in which the epoxides, popular for microscopy, can be cured, and exacerbates the extraction of soluble tissue components, unless the fixation regime has been extensive (i.e. aldehyde plus osmium post-fixation). The tissue's immunoreactivity is greatly reduced or lost by such fixation procedures, and so epoxy resin embedding should be avoided if possible.

   Acrylic resins can be polymerised by all of the methods of polymerisation listed above, of which heat is the most damaging to tissue treated with low fixative concentrations (e.g. 0.1-0.2% glutaraldehyde). Chemical, UV- and blue-light polymerisation can be carried out at low temperatures (0°C or lower) to improve retention of immunoreactivity and minimise extraction of cellular components.

   The cross-linking of acrylic resins is achieved by the presence of free radicals involved in addition reactions on polymerising chains (Carlemalm and Villiger, 1989). Usually the presence of a catalyst (or initiator) is required to start the polymerisation reaction. Free radicals can be produced or initiated by heat, the presence of peroxides or by light. For heat, dibenzoyl peroxide is added as a catalyst to the resin to initiate polymerisation. Chemical polymerisation at room or low temperatures is again initiated by the addition of dibenzoyl peroxide, but now requires the help of an accelerator/activator (an amine compound) to start the production of free radicals, as the activity of the catalyst (dibenzoyl peroxide) is linked to temperature. UV-light provides the energy to produce free radicals with the help of an initiator (an ethyl or methyl ether).

The amount of catalyst or initiator activated will determine the number of polymerising chains; the temperature at which the particular method of polymerisation is carried out will affect the rate of the reaction, as chain elongation is dependant upon temperature (Carlemalm and Villiger, 1989). Acrylic resin polymerisation is an exothermic reaction, and so the choice of the amounts of catalyst/initiator and accelerator/activator, the energy input from the UV-light source (wattage and distance of source from resin capsules; indirect vs. direct illumination), and volume of resin to be polymerised (preferably below 0.7 ml) will all affect the final cross-linked density of the resin and the rise in temperature that occurs during any of the polymerisation methods used. Indeed, the temperature rise for Lowicryl resins at -35°C with UV-light polymerisation is less than 2°C (Weibull, 1986), if (a) the volume of resin is small, (b) the distance of light source to resin capsules is adequate, (c) the UV-light source is diffuse (indirect), and (d) the cold cabinet allows for an efficient heat exchange. For all polymerisation temperatures, any heat rise caused by chemical polymerisation methods can be reduced by placing the resin filled gelatin capsules (again small volumes) in an aluminium block acting as a heat sink (Acetarin and Carlemalm, 1982; Newman and Hobot, 1987). Heat sinks can also be used for UV-light polymerisation (Sect. 5.4; Ashford et al, 1986; Glauert and Young, 1989; Weibull, 1986). Blue light, advocated for the polymerisation of LR Gold at -20°C (Sect. 2.2.2), involves the use of a strong heat source. It is, therefore, unsuitable for polymerising acrylic resins, for coupled to the already exothermic reaction occurring during polmerisation, the heat produced now would make it very doubtful that the temperature is maintained at, or even close to, -20°C.

Control of the cross-linked density of the resin is therefore possible with acrylics, bearing in mind the above points when preparing formulations. Minimising cross-linked density may be important in improving the sensitivity of the immuno system employed by allowing greater access of antisera to antigens (Chap. 10). The final cross-linked density of the resin is probably best controlled by chemical polymerisation, where the addition of catalyst and activator can be more finely balanced to dictate the rate of reaction, the degree of polymer chain cross-linking, and the temperature rise produced during the exothermic reaction. It should be borne in mind that tissue which has been stabilised with low fixative concentrations can contain active components (e.g. amino groups) which will aid chemical polymerisation (perhaps by increasing the rate of reaction and the heat produced), and become involved in the cross-linked polymer (Carlemalm and Villiger, 1989).

Reducing the temperature of polymerisation to 0°C alleviates this problem (Newman and Hobot, 1987; Sect. 3.3.1.2).

2   The Resins

2.1   Epoxy Resins

2.1.1 Introduction

Epoxy resins were tested quite early on in the development of EM embedding media as alternatives to methacrylate (Maaløe and Birch-Andersen, 1956; Glauert et al, 1956; Glauert and Glauert, 1958; Luft, 1961; Glauert, 1965, 1975). The epoxy resins gave uniform curing, and were found to be beam-stable. Uncrosslinked methacrylate is only two-dimensionally bonded and is rapidly destroyed in the electron beam. It is also prone to shrinkage, sometimes uneven, causing shearing and "polymerisation damage" (Pease, 1964; Glauert, 1965, 1975). By the time these problems were addressed the epoxy resins, which had no such problems, were firmly established. The history and development of epoxy resins and the other group of beam-stable, highly crosslinked resins, the polyesters, is well documented elsewhere (Glauert, 1965, 1975, 1991).

2.1.2 Choice of Resin

The basic formulation of most epoxides is the same. They consist of a mixture of base resin, hardener and accelerator. Flexibilisers and plasticisers may be added to improve sectioning quality. There are many proprietary preparations of these constituents, obtainable either as seperately purchasable items for experimental use or in the form of embedding kits for routine application (Agar Scientific; Fisons Scientific Equipment; Polysciences; Taab). They are all very powerful, three-dimensionally bonded resins chosen for embedding purposes from industrial formulations (e.g. adhesives, paint fillers). When cured, they are resistant to the heat and static electricity generated by an electron beam and are almost impervious to aqueous solutions, particularly if acid. As a result, highly alkaline solutions of toluidine blue are commonly used to stain semithin tissue sections, and similarly alkaline solutions of lead salts are used to stain thin sections. For other stains and immunolabelling, sections usually have to be etched to remove the resin and expose the tissue (Sect. 8.1.1).

It is important to bear in mind that epoxy resins are composed of several components, some of which may be carcinogenic or mutagenic agents. These can include not only the resin itself but the hardeners and flexibilisers as well. The amines used as accelerators are toxic (Causton, 1984; Causton et al, 1981). Acrylic resins do not have these properties, except when using amine accelerators for chemical polymerisation (Sect. 5.3), but can be allergenic. Therefore great care is required when handling all resins, always wearing appropriate gloves and working in a fume cupboard.

Araldite (based on the resin Araldite CY 212 made by Ciba-Geigy, Basel, Switzerland) may be regarded as typical of the group (Glauert, 1975). It is very viscous and is, therefore, difficult to measure out safely and mix for use. A useful formulation is :-

Araldite CY 212	Resin	48.0 g
Dodecenyl Succinic Anhydride	Hardener	44.0 g
Benzyldimethylamine (BDMA)	Accelerator	1.6 g

This formulation produces a reasonably hard block.
Glauert (1991) suggests a similar formulation which may produce an even harder block:-

Araldite CY 212	Resin	44.0 g
Dodecenyl Succinic Anhydride	Hardener	42.0 g
BDMA	Accelerator	2.6 g

Dibutyl Phthalate (plasticiser, 1.2 g) is only added to Araldite MY753 and 502 (Glauert, 1991).

Established laboratories will have different preferences as to hardness and other sectioning characteristics of the polymerised resin and will have developed their own formulations. Some caution is necessary, however, because there can be large batch to batch variation in the characteristics of the resins, and some of the additives, in particular amine accelerators such as BDMA, have relatively short shelf-lives often making reproducibility difficult. Exposure to moisture will shorten their shelf life.

Although Araldite is soluble in the common dehydrating agents, ethanol and acetone, intermediate solvents (Sect. 1.3.3) improve infiltration. Even so, long infiltration times are required. Warming, which reduces viscosity, may be necessary to infiltrate dense or large specimens. To achieve curing of the resin in a reasonable time (36-48 hours), heat (60°C) must be used, in addition to the presence of an amine accelerator.

The other epoxy resin in popular use is Epon. Originally manufactured as Epon 812 by Shell, it is now avaliable as Epon 'embedding' kits from the various EM suppliers. It is less viscous than Araldite, but is still difficult to measure out and mix for use. A useful formulation derived from Glauert (1991) is presented below. Further information to vary the hardness of Epon blocks can be found in the publication by Glauert (1991):-

Epon	Resin	48.0 g
Dodecenyl Succinic Anhydride	Hardener	18.0 g
Methyl Nadic Anhydride	Hardener	30.0 g
BDMA	Accelerator	2.8 g

A low viscosity formulation, Spurr's resin (Spurr, 1969; Kushida, 1971; Glauert, 1975; Agar Scientific; Fisons Scientific Equipment; Polysciences; Taab), designed to penetrate dense specimens more easily, is available. It is based on the epoxy resin ERL 4206, and is even more carcinogenic than the other epoxies. It should only be handled with gloves in a fume cupboard using extreme caution. A useful formulation is given below, and the hardness of the resin can be varied according to Spurr (1969):-

ERL 4206	Resin	20.0 g
DER 736*	Flexibiliser	12.0 g
Nonenyl Succinic Anhydride	Hardener	52.0 g
Dimethylaminoethanol	Accelerator	0.8 g

(*Diglycidyl ether of polypropylene glycol)

The only water miscible epoxy resin that survives is Durcupan (Stäubli, 1960; Glauert, 1975; Agar Scientific; Fisons Scientific Equipment; Taab). It should be noted that the water-soluble formulation of Durcupan is different from Durcupan ACM, which is Araldite M/CY 212 (Glauert, 1991). Tissue is dehydrated in a gradient of increasing concentrations of Durcupan in water thus avoiding the use of dehydration solvents (Sect. 1.3.3). Extraction may well be reduced, but there is a fixative effect from the resin (Gibbons, 1959) which prevents the tissue from being available for cytochemistry (Leduc and Bernhard, 1962) and probably, therefore, immunocytochemistry as well.

2.2 Acrylic Resins

Many of the problems associated with the three-dimensional bonding and hydrophobia of epoxides, where they are applied to immunocytochemistry, are avoided by using acrylic embedding resins. Their early difficulties of use for electron microscopy were overcome just before the epoxy resins swept them aside in popularity. The potential of this group of resins was, therefore, never fully realised. Acrylics can be made more resistant to electron beam damage by the addition of crosslinking agents such as divinyl benzine and styrene (Kushida, 1961a, 1961b), or by the addition of aromatic groups in to their structure. However, many purpose-built, beam-stable acrylics are now available, providing great scope for the microscopist to preserve both tissue structure and reactivity. The acrylic resins are not listed as carcinogenic, although the amines used as accelerators/activators (Sect. 1.3.4) are very toxic and so require careful handling. The resins may be allergenic and gloves and the fume cupboard should again be employed when handling them (Sect. 2.2.3.3, Varying the Hardness of the Resins). Those resins which have little resistance to the electron beam, for example the methacrylates in the absence of cross-linkers and stabilisers, will not be considered here.

2.2.1 LR White

2.2.1.1 Introduction

The physical and chemical properties of LR White (manufactured by the London Resin Company; also distributed by Agar Scientific; Fisons Scientific Equipment; Polysciences; Taab) give it great versatility as an embedding resin. Before polymerisation it is very free-flowing with a viscocity (10 cps) similar to water enabling it to penetrate tissue rapidly. It can be polymerised in different ways giving rise to a variety of techniques to suit different tissue requirements. Unlike some acrylics, the monomer is only partially water miscible but will polymerise in the presence of up to 12% by volume of water (B.Causton, London Resin Co., Basingstoke, Hants., UK, personal communication) enabling unosmicated tissue to be embedded while incompletely dehydrated (Newman et al, 1982, 1983a). This has the advantage of reducing the extraction caused by exposing unosmicated tissue to concentrated organic solvents. LR White has good sectioning properties and sections cut from it are hydrophilic and freely permeable to all aqueous solutions including those with a neutral or acid pH. Semithin sections of tissue in LR White can be stained in both acidic and basic dyes and immunolabelled with either colloidal gold or peroxidase/diaminobenzidine, although both these methods will require photochemical intensification. Thin sections are stable in the electron beam even when mounted unsupported on 300 mesh grids. All sections immunolabel and

Figure 3. Part of a kidney glomerulus from a kidney biopsy immersion fixed in 1% glutaraldehyde. (a) Post-osmicated, fully dehydrated and embedded in Araldite. (b) Not osmicated, partially dehydrated and embedded in LR White by the cold chemical catalytic method (Sect. 3.3.3). Both (a) and (b) were 'block-stained' in uranyl acetate and counterstained with only lead citrate. The structure of the LR White embedded kidney is as good as that of the osmicated Araldite embedded tissue. The LR White embedded kidney can also be immunolabelled (see Fig. 18). (Mag. x for both 22,000).

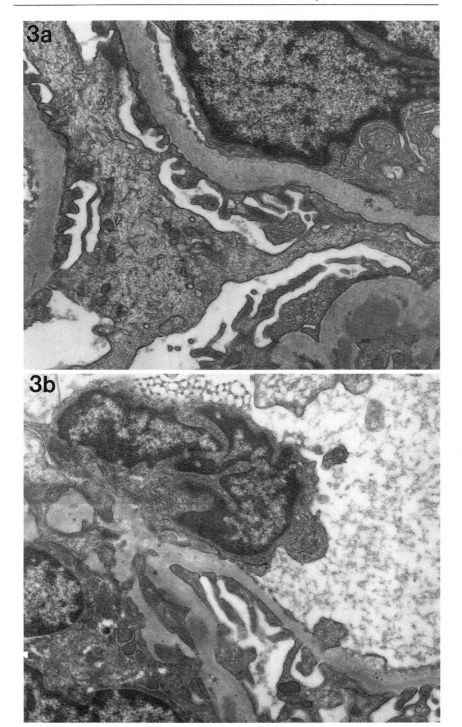

counterstain intensely without the need for etching (sodium ethoxide, hydrogen peroxide - Baskin et al, 1979), unmasking (sodium periodate - Bendayan and Zollinger, 1983), or protease digestion (Sect. 8.1).

2.2.1.2 Historical Perspective

The use of LR White for immunoelectron microscopy was first described by Newman et al (1982, 1983a) who introduced the partial dehydration of unosmicated tissue, prior to embedding, as a means of minimising extraction. When embedding unosmicated tissue in to LR White, partial dehydration markedly increases the tissue's sensitivity to antisera and lectins (Newman et al, 1982, 1983a), and produces differences in structural appearance from those seen in fully dehydrated tissue (Newman and Hobot, 1987). In addition, evidence for the retention of phospholipids in the nucleus after partial dehydration and LR White embedding has been furnished by cytochemical and histochemical studies (Zini et al, 1989). It enhances the periodic acid-Schiff (PAS) and periodic acid-thiocarbohydrazide-silver proteinate (PA-TCH-SP) reactions (Neiss, 1988) and a recent report suggests that increased immunosensitivity is also shown by osmicated tissue embedded in LR White after partial dehydration (Mutasa, 1989).

The first immunolabel reported to be used on LR White thin sections was immunoperoxidase/diaminobenzidine (DAB) (Newman et al, 1982, 1983a; Newman and Jasani, 1984a). The receptivity of tissue embedded in LR White for electron opaque stains and fixatives, made it necessary to use gold chloride to increase the electron density of the DAB, so as to avoid obscuring it by increasing the density of the surrounding tissue (Newman et al, 1983b). This can occur with the traditional method using osmium. Counterstaining with electron opaque stains was omitted.

Immunocolloidal gold was shown to be equally applicable to LR White sections (Newman and Jasani, 1984b) with the advantage that the good ultrastructure preserved, even in the absence of osmium post-fixation, could be demonstrated by light counterstaining without obscuring the colloidal gold marker. Both DAB and colloidal gold can be used on semithin LR White sections for light microscopy provided that they are photochemically enhanced. DAB is often either faint or completely invisible, particularly if the semithin section is less than 0.5 μm thick (Newman et al, 1983c, 1986) but can be turned black with for example Amersham's Silver Enhancement Kit (Amersham International). Colloidal gold, also usually invisible on semithin resin sections, catalyses the reduction of metallic silver from solutions of silver salts (Holgate et al, 1983; Danscher and Rytter-Norgaard, 1983; Springal et al, 1984; Bowdler et al, 1989) so that it too can be rendered black and easily visible with, for example, BioClin's "C-Gold" silver staining kit (BioClinical Services). Finally both labels have been shown to be compatible for high sensitivity double immunolabelling (Newman et al, 1986, 1989; Newman and Hobot, 1987).

Controlled chemical catalytic polymerisation of LR White was introduced in 1986 by Yoshimura et al as an alternative to heat polymerisation. The method described is very rapid and often increases immunosensitivity markedly, probably by reducing tissue extraction by greatly shortening the time over which it is infiltrated with resin and polymerised. Additionally, during polymerisation the peak temperature of 60°C is only sustained for 1 to 2 minutes before the block rapidly cools back to room temperature.

A modification was needed to prevent tissue distortion caused by too rapid a polymerisation of lightly fixed tissue (Sects. 1.3.4/3.3.1.2). The polymerisation was conducted at 0°C using a pre-cooled resin/catalyst mixture (Newman and Hobot, 1987). This method is less rapid but still greatly reduces the time for which the tissue is in contact with the resin monomer. When used in conjunction with partial dehydration not only is there an improvement in tissue reactivity but some structures thought only to be preserved by low temperature techniques (PLT), such as the trans-tubular Golgi apparatus (Roth et al, 1985) and the bacterial cell envelope (Hobot et al, 1984) now make their appearance (Newman and Hobot, 1987).

An important advantage in using the rapid embedding methods for LR White, introduced by both Yoshimura et al (1986) and Newman and Hobot (1987), lies in providing a fast diagnostic service. The versatility of the methods, in producing embedded tissue with good preservation of structure and immunoreactivity, is enhanced by the ease with which LR White can be exploited for either light or electron microscopical cytochemistry and immunocytochemistry (Bowdler et al, 1989; Sect. 1.3.1).

LR White, as an acrylic resin, possesses a low electron cross-scattering potential in relation to the embedded tissue, a property it shares with the Lowicryls (Carlemalm et al, 1985a; Sect. 2.2.3.2). Tissue that has, for example, been immunolabelled with peroxidase or double immunolabelled with colloidal gold and peroxidase, therefore requires little or no contrasting agents to allow for visualisation of the sections in the transmission electron microscope (Newman et al, 1983a, 1983b, 1986, 1989; Newman and Hobot, 1987). These advantages of LR White have also been applied, for the first time, to the differential staining with uranyl atoms of newly synthesised cell wall polymers in bacteria (Clarke-Sturman et al, 1989; Merad et al, 1989).

2.2.1.3 Versatility of LR White

LR White is available in three grades, hard, medium and soft. In practice, however, in common with many acrylic resins, the texture of the block and its sectioning characteristics can be varied by altering the extent to which the resin is polymerised. The original or hard-grade resin is the one preferred for immunocytochemistry because this resin can be considerably under polymerised and yet retain good sectioning qualities. By reducing the cross-linked density of the resin polymer (i.e. under-polymerising, Sect. 1.3.4), it may be possible to improve the immunosensitivity of tissue sections, by, for example, allowing easier access for antibodies in to sections (discussed in Chap. 10). Thin sections cut from blocks which have been deliberately under-polymerised may be delicate and should be treated with care whilst immunolabelling. Their polymerisation will be completed by the electron beam during EM viewing, provided a few seconds of time is allowed for them to equilibrate at low magnification and with the beam well spread. Should they still prove unstable thin sections can be mounted on support films, or given a light carbon coating, or alternatively, the resin blocks may be further polymerised using gentle heat (Chap. 5). Later descriptions of the use of LR White for resin embedding and immunocytochemistry will always refer to the hard-grade, as originally sold by the manufacturers complete with added catalyst.

Until recently, LR White could only be purchased with chemical catalysts included. To all intents and purposes, therefore, the resin had already begun to polymerise on arrival in the laboratory. For the kind of critical polymerisation advocated for high sensitivity immunolabelling, such resin has a useful shelf-life of only about 3 months even when properly stored in a refrigerator at 4°C. However, the resin mixture is now available without catalysts (ostensibly for hot climates). The manufacturers provide benzoyl peroxide/catalyst reagent separately, to be added and mixed with the resin as required ('Uncatalysed LR White, Sect. 5.5). This gives the user far more control over the extent of resin cross-linking generated during polymerisation (Sect. 5.3).

Polymerisation can be achieved by at least three methods: heat, chemical catalysis or UV-light (Chap. 5). The method chosen will largely depend on the extent of tissue fixation tolerated by the antigen under investigation. Heavily fixed tissue, for example where large blocks (> 3 mm^3; N.B. thickness not volume is important) are necessary or if the tissue is osmicated and, therefore, more dense, will need lengthy infiltration and should be heat polymerised. Smaller blocks of delicately preserved tissue better retain their immunoreactivity and structural integrity if polymerised more rapidly with chemical catalysts in the cold. Provided the blocks of tissue are unpigmented and very small (< 0.5 mm^3) they can also be polymerised by UV-light. Both chemical catalysis and UV-light can be useful for polymerisation at low temperature. However, LR White freezes at about -27°C, so low temperature polymerisation following, for example, PLT (or even lower temperature polymerisation following cryosubstitution or freeze-drying) is best dealt with by the Lowicryls.

Figure 4. Kidney tubule of rat perfusion-fixed with 0.5% glutaraldehyde and, following partial dehydration, embedded in LR White by the cold chemical catalytic method (Sect. 3.3.3). The transtubular Golgi apparatus is clearly seen although it was thought that this could only be shown following PLT. The section was labelled with biotinylated e-pha lectin (*Phaeseolus vulgaris*) and avidin colloidal gold (Sect. 7.2.3.1), which has localised a lysosomal body adjacent to the Golgi (see Figs. 7 & 9). (Mag. x 34,000).

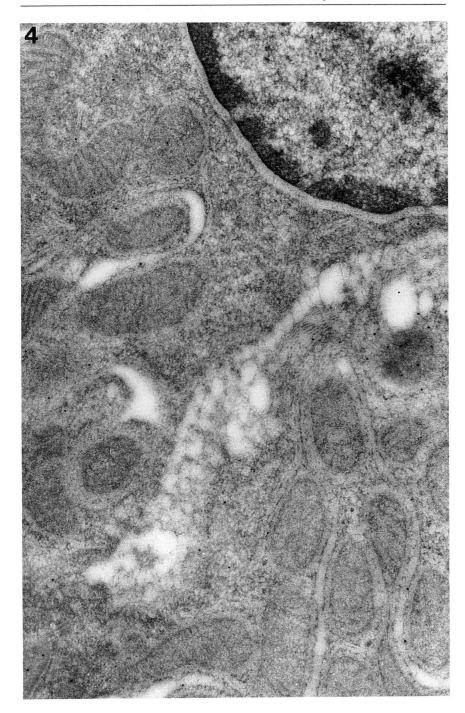

2.2.2 LR Gold

The formulation for the acrylic resin LR Gold (London Resin Company; Agar Scientific; Fisons Scientific Equipment; Polysciences; Taab) is very similar to that of LR White (B.Causton personal communication) although the composition of either remains unpublished in the literature. The physico-chemical properties that make LR White so suitable for cytochemisty and immunocytochemistry are, therefore, mirrored in LR Gold, which can be used in precisely the same way. Recently, however, Taab Ltd. have made available the composition of LR Gold (LR Gold Safety Data Sheet, Taab Ltd.). Interestingly, the aromatic backbone that gives LR Gold its increased resistance to the electron beam, bisphenol A, can also also be used to produce epoxy resins. The amount of the accelerator/activator dimethylparatoluidine added to LR Gold formulations (0.4%; manufacturer's recommendation) is far in excess of the amounts required to chemically polymerise LR White (Sect. 5.3.1) or the Lowicryls (Sect. 5.3.2). In these amounts it will give an unacceptable heat rise during the exothermic polymerisation reaction (see also below, this section) if the recommended amount of benzoyl peroxide is added (0.9%). It is important to note the source of benzoyl peroxide being used (refer to points discussed in Sects. 5.1 and 5.5). The amounts of catalyst and accelerator added should be similar to those suggested for the Lowicryls (Sect. 5.3.2) or 'Uncatalysed' LR White (Sect. 5.5). A heat sink must be employed (Sect. 5.1). Alternatively, a photoinitiator (benzil, 0.1%) can be added to LR Gold to make it sensitive to blue light for the purpose of polymerisation.

LR Gold has been advocated, by the manufacturers, for embedding "unfixed" tissue at low temperature in order to retain histochemical reactivity in enzymes which are denatured by aldehyde fixation. Fresh tissue is placed in to methanol at 0°C (two changes 30 min each) and when dehydrated the temperature is lowered to -20°C. The tissue is infiltrated at -20°C with LR Gold resin (1 hour at first, then overnight, a third change for 1 hour the following morning) before final embedding and polymerisation at -20°C with a lamp emitting blue light (any low wattage microscope bulb). Little data has been published using this method and there are obvious problems. Tissue placed in methanol, even at 0°C, is almost certainly "fixed" at least to some extent. The lack of tissue stabilisation, imparted through chemical cross-linking during aldehyde fixation, can also lead to structural alterations caused by the dehydrating solvent. Blue light involves the use of a strong heat source, and, as with all acrylics, LR Gold polymerises exothermically which also means the production of heat, so it is doubtful whether the temperature is accurately maintainable at -20°C (Sect. 1.3.4). The ultrastructure of tissue preserved by this method is very poor.

LR Gold has been used to embed aldehyde fixed tissue at -20°C for immunocytochemistry and cytochemistry. The value of this method, over room temperature and 0°C methods, is again doubtful. Polymerisation almost certainly does not take place at -20°C for reasons already given and even if it did, it has not been shown that it leads to any improvement in ultrastructural or antigenic preservation. Published work with LR Gold has shown that the level of immunolabelling is the same as that obtained when using LR White (Berryman and Rodewald, 1990; Takada et al, 1992). The only improvement that was observed with LR Gold over LR White was in the preservation of lipid-rich structures in the rat nephron (Takada et al, 1992). Indeed LR White can be used in place of LR Gold at -20°C (Sects. 3.4/5.3.1.2/5.4.2.2; Berryman and Rodewald, 1990; Takada et al, 1992). Like LR White, LR Gold can be polymerised using UV-light (Sect. 5.4) which provides more hope of keeping the polymerisation temperature nearer to -20°C, but both resins freeze at about -27°C so they cannot be used at the sub -30°C temperatures advocated for the progressive lowering of temperature method devised for the Lowicryls.

2.2.3 The Lowicryls

2.2.3.1 Introduction

Singly the most important property of this group of acrylic resins is their low viscosity at temperatures of -35°C and below. They can be polymerised with ultra-violet (UV) light whilst still at low or very low temperature. As a result, their most popular applications have been in the use of low temperature to preserve tissue structure and cytochemical and immunocytochemical reactivity. In this capacity they usefully extend in sensitivity the series of methods begun with LR White.

The past decade has seen an increase in the use of the Lowicryl resins (K4M, HM20, K11M and HM23 - manufactured by Polysciences; also distributed by Agar Scientific; Fisons Scientific Equipment; Taab), especially that of Lowicryl K4M, for applications in immunoelectron microscopy. Although this has been beneficial in our understanding of cellular processes, it also has meant that some of the advantages of the Lowicryls and their original aims have been temporarily overlooked. There is now renewed interest in the other Lowicryls, especially Lowicryls HM20 and HM23, as it is realised that these resins are also useful in immunoelectron microscopy. This is to be welcomed, but greater advances are more likely to be made if the Lowicryl system is fully understood. Some details of their chemical composition can be found in either Carlemalm at al (1982a) for Lowicryls K4M and HM20, or in Acetarin et al (1986) for Lowicryls K11M and HM23.

2.2.3.2 Historical Perspective

The technical problems once associated with the use of acrylics have largely been overcome and so the way has been paved for the introduction of the Lowicryls (Kellenberger et al, 1980). Fernández-Morán (1960, 1961) showed early on that low temperature processing was important in preserving ultrastructure, lowering the degree of reorganisation and denaturation of molecular structure that can occur above 0°C. Besides employing rapid freezing and cryosubstitution to illustrate the advantages of using low temperatures, Fernández-Morán (1961) embedded fixed tissue in acrylic resin following dehydration in an organic solvent at "progressively lower temperatures".

Later Douzou et al (1975), Douzou (1977) and Fink and Ahmed (1976) demonstrated that enzymes could maintain their native structure and activity at -50°C to -100°C in 70-90% organic solvent. This is probably because at low temperature the amplitude of molecular thermal vibrations is lowered, effectively altering the dielectric constant of the organic solvent (Artymink et al, 1979; Frauenfelder et al, 1979). Temperatures above -30°C may not be low enough to be effective (MacKenzie, 1972). A basis was, therefore, provided for the re-introduction and improvement of the Progressive Lowering of Temperature (PLT) method (Kellenberger et al, 1980; Carlemalm et al, 1980, 1982a) whereby an increase in organic solvent concentration is combined with a progressive decrease in temperature, followed by infiltration with, and polymerisation of, resin (Lowicryl) at low temperature.

The Lowicryl resins have been constructed to provide differences in polarity and hydrophilia. Lowicryl K4M directly tolerates up to 10% by weight of water, whereas Lowicryl HM20 can accommodate only 0.5%. Like LR White, Lowicryl K4M will combine with 70% ethanol in the ratio of 2 parts resin to one part 70% ethanol (Hobot, 1990; Hobot and Newman, 1991), but Lowicryl HM20 is only miscible with 90% ethanol, although at a 1:1 ratio. The later additions to the Lowicryl series, Lowicryls K11M and HM23 (Acetarin et al, 1986), fit along this sliding scale of polarity (more polar > less polar, i.e Lowicryls K4M > K11M > LR White > Lowicryls HM20 > HM23; Hobot and Newman, 1991). These properties may have significant implications in the preservation of, for example, the hydration shells around proteins which may be better retained by the less polar resins (Kellenberger et al, 1980). UV-light polymerisation at low temperature occurs without serious heat rise (Weibull, 1986) so that a true low temperature embedding regime can be formulated to fit the nature of the proposed tissue specimen. In 1980, Weibull et al showed that low temperature embedding resulted in an improved ultrastructure of the thylakoid membranes of pea or spinach chloroplasts. Reports on other tissues followed (Kellenberger et al, 1980; Roth et al, 1981; Carlemalm et al, 1982a; Armbruster et al, 1982), together with showing the advantages of the Lowicryls for immunocytochemistry (Armbruster et al, 1984; Roth et al, 1981).

First systematic results with the Lowicryls to show the advantages of low temperature embedding for structural preservation were obtained using protein crystals as models (Carlemalm et al, 1982a). It was shown that by choosing a dehydrating solvent with a polarity similar to that of the resin an improvement could generally be seen in the crystal's molecular order when embedded at low temperature (-35°C to -50°C). This improvement was not shown when the crystals were processed and embedded at higher temperatures (0°C or 20°C). Following this, two new proposals for cellular structure involving PLT methods have been put forward. A new model for the organisation of the gram-negative bacterial cell envelope, the periplasmic gel, was suggested in part by results from PLT (Hobot et al, 1984) and the model was supported by direct observations on frozen-hydrated sections of unfixed rapidly frozen bacteria (Dubochet et al, 1983). In the second development, the arrangement of the Golgi apparatus in to an extensive trans-tubular network continuous with the Golgi stacks was described in liver of rat embedded in to Lowicryl K4M by the PLT method (Roth et al, 1985). These new structural features have also been observed in tissue embedded in LR White by a rapid, room temperature, partial dehydration technique (Newman and Hobot, 1987). Recently, for tissue embedded in Lowicryl K4M in a similar way, the bacterial cell wall findings have again been reproduced (Hobot, 1990). These observations cannot be made in post-osmicated, epoxide-embedded tissue or in aldehyde fixed cells that have been fully dehydrated and embedded in an acrylic resin at room temperature or 0°C (Hobot et al, 1984; Newman and Hobot, 1987). The ultrastructural observations complement the results obtained in retaining similar levels of immunoreactivity within tissue for the detection of sensitve antigenic sites when using low concentrations of glutaraldehyde and either partial dehydration (with LR White or Lowicryl K4M) or PLT (with Lowicryl K4M) techniques (Hobot and Newman, 1991; Newman and Hobot, 1989; Sect. 1.3.2.1). An advantage of using the low temperature procedure of PLT lies in the ability of the technique to retain not only sensitive or low concentration antigenic sites by minimising extraction, but also to preserve ultrastructure at a higher level when using low amounts of glutaraldehyde for short fixation times (0.1%/15 min; Hobot, 1989; Newman and Hobot, 1989).

Acrylic resins also have the advantage of low composite density in comparison with the higher density of biological tissue. Epoxides and polyesters have a high electron cross-scattering potential which is similar to that of the tissue, and as a result, section contrast is poor and the tissue must be heavily counterstained with electron opaque heavy metals (Carlemalm et al, 1985a). The Lowicryls and LR White do not require such extensive counterstaining for visualisation following immunolabelling, especially after applying double immunolabelling techniques (Sects. 2.2.1.2/7.4). The higher density (or atomic number, Z) of biological tissue in Lowicryl can be visualised by scanning transmission electron microscopy (STEM) which uses the differences in atomic number and the ratio of elastically to inelastically scattered electrons to create an image of biological structures, unaltered by the addition of heavy metals. The contrast is due solely to the tissue itself: atomic- or Z-contrast (Carlemalm and Kellenberger, 1982; Carlemalm et al, 1985a). Imaging of this kind has made it possible to visualise the transmembrane proteins of the septate junction in *Drosophila melanogaster* testis (Carlemalm and Kellenberger, 1982; Garavito et al, 1982) and contributed towards a deeper understanding of the structure of the bacterial cell envelope (Hobot et al, 1981,

1984), such that the prospects for achieving more information concerning ultrastructure are now improved (Carlemalm et al, 1982b).

As well as preserving the conformation of proteins, PLT methods also reduce the extraction of soluble tissue components. This is illustrated by the way in which it is possible to retain lipid after PLT employing ethanol dehydration and Lowicryl embedding, especially with Lowicryl HM20 (Weibull et al, 1983; Weibull and Christiansson, 1986). The extraction of radiolabelled lipid from the cytoplasmic membrane of the bacterium *Acholeplasma laidlawii*, fixed initially in glutaraldehyde, occurs to the extent of 64% following osmium and uranium post-fixation and routine processing in to Epon or Lowicryl K4M. With Lowicryl HM20 the loss is 22%. In the absence of osmium and uranium post-fixation, the loss is less than 50% in either Lowicryls K4M or HM20 after PLT at -35°C (N.B. dehydration finished at 90%; Weibull et al, 1983). Greater retention is possible with Lowicryl HM20 which, unlike Lowicryl K4M, can be used down to -50°C. At this very low temperature up to 89% of lipid was retained (Weibull and Christiansson, 1986). It should be borne in mind, that the lipids from *A. laidlawii* are unusual in their composition. But, all the lipids in this organism are present in the cytoplasmic membrane and can be selectively radiolabelled by varying the growth medium composition. Thus, the incorporation and extraction of lipids from *A. laidlawii* can be monitored accurately, and is a distinct advantage in this model system in comparison to others. However, the results from *A. laidlawii* compare favourably with data from other systems. After glutaraldehyde fixation (1%, lower concentrations were not studied) and PLT embedding in Lowicryl HM20 at -50°C, similar figures have been obtained for phospholipid retention in erythrocyte ghosts (84%); and values for chlorophyll retention in spinach chloroplasts and thylakoids were 60% and 34% respectively (Weibull and Christiansson, 1986).

Very importantly, Weibull and Christiansson (1986) showed that the extent of lipid extraction from *A. laidlawii* at -50°C depended on the dehydration medium used (ethanol for the results above), with acetone allowing a retention of 86% of the lipid and methanol only 54%. Similarly for *A. laidlawii* cells, prepared by cryosubstitution and embeding in Lowicryl HM23 at -70°C, lipid retention values of 95% for acetone and 55-85% for methanol were found, substitution in these organic solvents being performed at -90°C. The amount of lipid extracted by the resin (Lowicryl HM23) at -70°C was neglegible (Weibull et al, 1984).

The addition of uranyl acetate either as a post-fixation step (Weibull et al, 1983) or for cryosubstitution (Humbel et al, 1983; Voorhout et al, 1991) further reduces lipid loss. It also does not seem to reduce the level of retention of immunoreactivitywithin tissue (Berryman and Rodewald, 1990; Erickson et al, 1987; Schwarz and Humbel, 1989; Sect. 8.1.6), except where nuclear material is to

Figure 5. 1% glutaraldehyde perfusion-fixed pancreas of rat embedded in Lowicryl HM20 following PLT (Sect. 3.5.3). The section is labelled for rat anionic trypsin with goat anti-rabbit IgG conjugated to 10 nm colloidal gold and counterstained with uranyl acetate and lead citrate. (Mag. x 22,400).

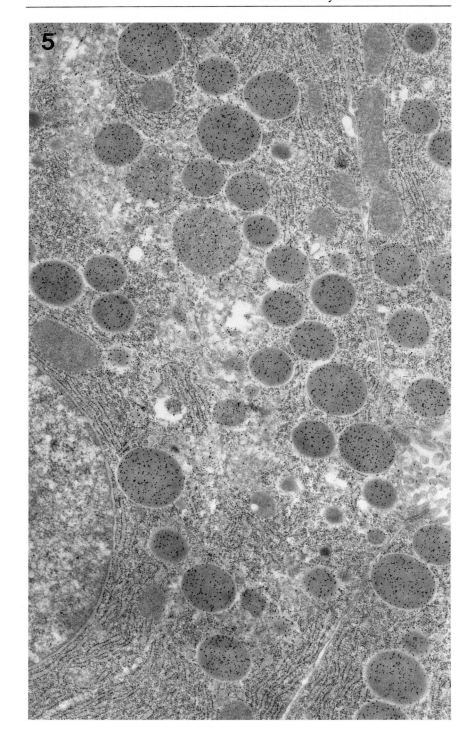

immunolocalised (J.A. Hobot, unpublished results). Adding 0.5% uranyl acetate to methanol during cryosubstitution can reduce lipid loss to 2% at -70°C, and to 4% at -30°C (Humbel, 1984).

2.2.3.3 Versatility of the Lowicryls

Past immunocytochemical studies have almost entirely been completed with Lowicryl K4M, with few results for Lowicryl HM20 (Lin and Langenberg, 1983, 1984; Bendayan et al, 1987), yet in terms of the ultrastructural preservation of the bacterial cell envelope Lowicryl HM20 gave results which were the same as those from Lowicryl K4M (Hobot et al, 1981, 1984), and indeed from LR White (Sect. 2.2.1.2; Newman and Hobot, 1987). The initial results to demonstrate the applicability of the Lowicryl resins generally showed improved structural preservation of membrane systems in various organisms following PLT and embedding in Lowicryl HM20 (Weibull et al, 1980; Kellenberger et al, 1980; Hobot et al, 1981; Armbruster et al, 1982; Carlemalm and Kellenberger, 1982; Garavito et al, 1982). Lowicryl HM20 has the additional advantages of lower viscosity, lower freezing point enabling it to be used down to -50°C, and good sectioning properties. It has an even lower electron cross-scattering density than Lowicryl K4M and, therefore, produces higher Z-contrast in STEM mode (Carlemalm et al, 1985a). Lowicryls K11M and HM23 have been introduced more recently (Acetarin and Carlemalm, 1985; Acetarin et al, 1986) specifically for cryosubstitution techniques where at -60°C (K11M) and -80°C (HM23) they are still of low viscosity and can be polymerised with UV-light.

The unpopularity of Lowicryl HM20 may be due to its having been classed early on as a "hydrophobic, apolar" resin causing it to be grouped in with the polyesters and epoxides. In practice, unlike the polyesters and epoxides, Lowicryl HM20 is miscible with a small amount of water (0.5% by weight) and will mix with 90% ethanol in a 1:1 ratio and therefore, whilst it may be more hydrophobic than Lowicryl K4M, it certainly should not be included with the truly apolar polyesters and epoxides. Lowicryl K4M would appear to be the most hydrophilic of the beam stable acrylic resins with a scale of diminishing polarity through Lowicryl K11M, (LR Gold, LR White) and Lowicryl HM20 to the least hydrophilic resin, Lowicryl HM23 (Sect. 2.2.3.2). Initial comparisons of immunolabelling between Lowicryl K4M (the most polar resin) and Epon (a non-polar resin), showed that the largest gain came from the reduction of unspecific background (Roth et al, 1981). Suspicions that Lowicryl HM20 was hydrophobic were supported by the high non-specific backgrounds that it also often developed when used for immunolabelling. The proposition that certain serum components may have an affinity for apolar resin surfaces (Dürrenberger, 1989) has given way to the less complex possibility that such surfaces have an unspecific attraction for basic protein, which can be neutralised by using large non-immune proteins i.e. bovine (ovine) serum albumin (Bendayan et al, 1987; Hobot, 1989) or gelatin (Dürrenberger, 1989). This simple expedient appears to allow Lowicryl HM20 and the other less polar resin, Lowicryl HM23, to be used as effectively for immunolabelling as Lowicryls K4M and K11M; and Lowicryl HM20 is now being more widely applied (Dürrenberger et al, 1988; Hobot, 1989; Hobot and Rogers, 1991; Schwarz and Humbel, 1989; Voorhout et al, 1991). The immunolabelling response obtained from sections of Lowicryl HM20 is the same as that for Lowicryl K4M (Hobot, 1989; Hobot and Rogers, 1991; Schwarz and Humbel, 1989). And both for light microscopical (basic) stains and peroxidase reactions it is in fact important to use Lowicryls HM20 and HM23 rather than Lowicryls K4M and K11M (Sect. 7.1).

Another choice that can be made with the Lowicryls, as also with LR White, is the way by which the resin can be polymerised (see Chap. 5): heat, chemical catalytic methods, or UV-light. This demonstrates that although the Lowicryls have in the main been used for low temperature embedding, the enormous versatility of the group allows methods of embedding from 65°C to -80°C. There is, therefore, a sound basis for a programme of study proposed by Kellenberger et al (1980) on the effects of various parameters on resin embedding which affect both ultrastructure and the retention of immunoreactivity within tissue, (studied in: Hobot, 1989; Hobot and Newman, 1991; Hobot et al, 1984), and which forms the experimental basis for the strategic approach presented in this book.

Lowicryls K4M and HM20
Lowicryls K4M and HM20 are supplied in kit form (Agar Scientific; Fisons Scientific Equipment; Polysciences; Taab). Each kit contains the relevant monomer for Lowicryl K4M or Lowicryl HM20 together with a cross-linker and an initiator, benzoin methyl ether, which allows polymerisation by UV light at low temperature (-35°C to -50°C). The cross-linker and initiator are the same for either kit (Carlemalm et al, 1982a).
Varying the Hardness of the Resins
The hardness of the polymerised resin block is dependant on the ratio of cross-linker to monomer. The more cross-linker the harder the block. When using K4M 13.5% by weight of cross-linker is recommended but the amount can vary from 4% to 20%. For HM20 15% is usually preferred but as with K4M the amount of cross-linker used can be varied from 5% to 17% by weight to suit individual requirements. Blocks with good general characteristics and intermediate hardness can be prepared by adding 1.35 g of cross-linker to 8.65 g of K4M monomer or 1.5 g of crosslinker to 8.5 g of HM20 monomer. These formulae will form the basis of the resins described below to which are added different catalysts/initiators and activators depending on the method and temperature of polymerisation. The various resin components should be weighed accurately in to a glass container. As with all acrylics, the method of mixing is important. Oxygen inhibits polymerisation so that once the resin monomer, cross-linker, etc. have been weighed out they should be mixed together thoroughly, using any method that excludes air such as bubbling dry nitrogen through the solution or gently stirring it with a glass rod. The

operation is carried out in a fume cupboard as some of the resin components are volatile. Gloves must be worn to avoid skin contact, as eczema can occur in sensitive individuals or after prolonged use (N.B. the articles by Tobler and Freiburghaus, 1990, 1991; Tobler et al, 1990). The wearing of the '4H glove' (Safety 4 A/S) is recommended (Tobler and Freiburghaus, 1991).

It is important that for all infiltration steps involving resin, the resin mixture of cross-linker and monomer contains either the appropriate UV-light initiator (Sect. 5.4.2) or dibenzoyl peroxide for heat or chemical catalytic polymerisation (Sects. 5.2/5.3).

Lowicryls K11M and HM23
Lowicryls K11M and HM23 come in kits containing monomer, crosslinker and two photoinitiators (labelled "C" and "J") for UV-light polymerisation.
Varying the Hardness of the Resins
To make 20 g of K11M, add 1 g of cross-linker to 19 g of monomer. To make 20 g of Lowicryl HM23, add 1.1 g crosslinker to 18.9 g monomer. Small changes in the proportion of cross-linker to monomer can result in considerable differences in hardness (Acetarin et al, 1986). Acetarin et al (1986) recommend that the resins are dried over molecular sieve (Linde type with a 4 Å pore) and that condensation is avoided at the low temperatures at which these resins are used.

Once again, for all infiltration steps involving resin, the resin mixture of cross-linker and monomer should contain the appropriate UV-light initiator (Sect. 5.4.2).

Rapid Embedding Procedures for Lowicryls
Rapid embedding would be most advantageous for diagnostic immunoelectron microscopy. Three methods for rapid embedding using Lowicryl K4M have been published. The first, a 4 hour technique (Altman et al, 1984), employs dehydration and infiltration at room temperature after the tissue has been heavily fixed in 3% glutaraldehyde and 3% formaldehyde and thoroughly washed. Polymerisation is achieved by direct UV-light at 0°C for 1 hour or less. The second (Simon et al, 1987) takes 48 hours at -10°C and is only a little shorter than PLT itself. Immunolabelling, because of the effects of heavily cross-linking tissue with high concentrations of glutaraldehyde may be poor (Sect. 1.3.2).

These two methods do not employ partial dehydration and all of its attendant advantages which are found in the third published method by Hobot (1990). The latter procedure takes 24 hours, but can be shortened to the same times found for LR White (Newman and Hobot, 1987; Yoshimura et al, 1986). It is possible, with fixation times of up to 1 hour and short buffer washes, to have the complete procedure over in 4 hours following the protocols set out for rapid embedding in either LR White or Lowicryl K4M (Sect. 3.3.2).

Lowicryl HM20 cannot be used for partial dehydration techniques, but it can be processed rapidly following protocols similar to those for LR White or Lowicryl K4M (Sect. 3.3.2). Dehydration can be either up to 90% or 100% organic solvent. When embedding in Lowicryl HM20 from 90% ethanol, the blocks are placed in to a 1:1 mixture of resin and 90% ethanol for 30 min, then four 20 min changes of pure resin. Polymerisation can be at room or cold (0°C) temperature by UV-light or chemical catalytic methods (Chap. 5).

3 Resin Embedding Protocols for Chemically Fixed Tissue

3.1 Tissue Handling

It is important to plan protocols carefully starting with a consideration of the kinds of tissue that are likely to be involved in the study and ways of handling them. They will probably fall in to one of two main types (Table 1) - 'free-living' cells (and cell-fractions) or 'solid' tissue (and cell-fractions).

3.1.1 Free-Living Cells (and Cell-Fractions)

Free-living cells may take the form of naturally occurring organisms in their native environment, cultured tissue in culture medium or various cells in body fluids. Cell types can be separated out one from another on the basis of size or weight using density gradients made up from dextran or using one of the proprietary differential gradients such as Ficoll, Ficoll-Paque, Metrizamide or Percoll (Sigma; Pharmacia, etc.). Cells may also be separated by size, density or following tagging with fluorochromes using a Fax Cell Sorter or Flow Cytometry (Beckton-Dickinson; Coulter Co.; Ortho Diagnostic Systems). Cell-fractions are produced from the centrifugation of macerated solid tissue or free-living cells through a differential gradient. Various organelle-containing fractions can be separated out according to the kind of density gradient and centrifugation speeds employed. Both cells and cell-fractions may be concentrated by simple centifugation or filtering methods and will need to be resuspended in non-proteinacious balanced saline which will sustain them, temporarily, without change, until they can be fixed. Where monolayers have formed, the culture medium is removed and the cells washed gently in balanced saline. A suspension can be achieved mechanically by scraping them up gently with a rubber 'policeman' or by chemically releasing them using enzymes such as trypsin and collagenase. Care must be excersised or the cells will be damaged by the releasing method. In addition, the orientation of the culture is lost, which can be a serious drawback when studying cell surface phenomena such as receptor sites. The cell or cell-fraction suspension is centrifuged to a pellet which can be immersion fixed directly in the centrifuge tube (Eppendorf bench microfuge using Eppendorf microcentrifuge tubes, or other as is convenient).

Free-Living Cells

Naturally Ocurring Cells
- Bacteria
- Yeast
- Sea-water organisms etc.

Cell Cultures
- Monolayers - (on 'Thermanox')
- Cell Suspensions

Body Fluids
- Blood
- Urine
- Lymph
- Cerebrospinal
- Coelomic etc.

Cell Fractions
- Maceration of above

Separation of Cells and Fractions:-
- Centrifugation through density gradients:-
  - Dextrans
  - Ficoll
  - Ficoll-Paque
  - Metrizamide
  - Percoll

Separation of Cells
- Fax Cell Sorting by:-
  - Size
  - Density
  - Fluorochromes

Fix:-
- Direct   -centrifuge, embed in agar etc.
            or process on Thermanox
- Pellets   -treat as solid tissue
- Filtrates -treat as solid tissue

Solid Tissue

Clinical
- Post-mortem
- Biopsy/Surgical

Fix:-
- Immersion

Animal
- Whole
- Organ
- Other/Biopsy/Surgical

Fix:-
- Perfusion
- Parenchymal perfusion
- Immersion

Cell Fractions
- Maceration of above

Separation of Fractions:-
- Centrifugation through density gradients as for free-living cells

Fix:-
- Direct   -centrifuge or embed in agar etc.
- Pellets   -treat as solid tissue
- Filtrates -treat as solid tissue

Table 1. Summary of the two main types of tissue and of the possible methods for their harvesting or collection and fixation.

The natural adhesives of cells and fractions will stick them together and keep the pellet whole following fixation, after which it can be treated as though it were a solid tissue block. Fixation will occur by diffusion and for this reason pellets should be kept small. For example, between 500,000 and a 1,000,000 cells will usually produce a 0.5 to 1 mm^3 pellet. Smaller pellets will be more difficult to handle but may be better fixed. The pellet should be eased away from the sides of the centrifuge tube after about 30 minutes of fixation so that it is free-floating in the fixative which can then penetrate from all sides. Gentle rotation on a rotary device will aid fixation. Because immersion fixation depends on diffusion, the fixative concentration must be high (> 0.5% glutaraldehyde or 2% formaldehyde or a mixture of both - see Sect. 1.3.2.1) and even then fixation will not be even. The outside of the pellet will often be much more heavily cross-linked than the centre.

To obtain an even level of fixation or to be able to experiment with different concentrations of fixative or times of fixation, pelleted cells or fractions must be resuspended in the fixative. In some cases, where the cells are surrounded by a medium with a good buffering capacity, the fixative can be added to the cell preparation. For example, fixatives may be added directly to cells in sea-water or to bacteria or other unicellular organisms or cells in culture media containing low levels of dissolved protein. The direct addition of the fixative to the cell suspension or preparation eliminates the risks involved in cell damage, which can occur during harvesting and concentration of fresh cells. Consideration should be given to the harvesting and concentration procedures where they are employed on unfixed cells and fractions. Centrifugation can induce anaerobic conditions within cells and cause ionic leakage, which can be reduced by using filtration techniques to collect or concentrate the cells (Epstein and Schultz, 1965). For cells dependant upon oxygen for their structural and antigenic integrity, filtration, under the correct temperature and aeration conditions to reduce ionic leakage, would be preferable (Moncany, 1982).

When fixation is completed, the suspensions have to be recentrifuged so that the fixative can be decanted and replaced with a washing solution. Alternatively, the cells can be filtered and washed directly on the filter. However, for preparations fixed directly in culture or solution, the cells and fractions will have lost their natural self-adhesive properties and will have to be centrifuged before every solution change. This procedure can be very tedious and time consuming and often leads to serious losses of material during changes. One possibility is to encase the tissue in a matrix such as agar, agarose, bovine serum albumin or gelatin. This technique is also important for bacterial cultures and where there is insufficient material from cell fractionation or cells from culture to form a manageable pellet.

The cell suspension can be harvested or concentrated by centrifugation or filtration and the cells resuspended in agar or in a low melting point agarose or gelatin mixture, dissolved by heating in distilled water or in an appropriate buffer/salt solution at a concentration of 1-4% w/v (Hobot et al, 1985; Hobot, 1991; Hobot and Newman, 1990). If the concentration is lower it may not allow for the formation of stable blocks. The heated solution is cooled and held to just above its gelling temperature in a water-bath or other suitable temperature controller and is mixed with the cell or cell-fraction preparation. The still liquid mixture is quickly taken up in to a silicon tube with a very small bore, allowed to solidify, pushed out of the tube, and cut in to small blocks. These can be processed further as if they were solid tissue blocks (Hobot et al, 1984; Kellenberger et al, 1972). Care must be taken to avoid diluting the sample too much when adding the agar or gelatin. An alternative technique, to reduce the dangers of diluting out very small samples, has been described by Hernández Mariné (1992). Briefly, a thin film of molten agar is spread on to a slide, and very small bubbles allowed to develop on the surface. An appropriately sized bubble is opened, an aliquot of sample is added and the bubble sealed with more molten agar. The solidified agar containing just the bubble of sample is carefully removed and procesed just like a tissue block (Hernández Mariné, 1992).

3.1.2 Monolayers

The orientation of a monolayer, which is lost by pelleting, can be preserved by fixing and processing the cells *in situ*. For convenience, the cells should be cultured on quick-release, plastic tissue-culture cover-slips (ICN-Flow) which are resistant to organic solvents so that the prosessing schedule, including dehydration, infiltration and embedding, can be completed by passing them, complete with cells, through the various solutions (Wynford-Thomas et al, 1986). This method allows easy experimentation with different fixatives, concentrations of fixative and times of fixation. The culture medium is washed away with balanced saline and each cover-slip, or, if they are cut up, each piece of cover-slip, can then be treated in a different way. Cells in monolayers are very rapidly cross-linked by neutrally buffered glutaraldehyde so low concentrations (< 0.5%) for short times (< 15 min) would suffice. After the glutaraldehyde is washed away (three 10 minute changes of buffer), the cells on the cover-slips can be either fully or partially dehydrated (see Sects. 3.2/3.3) and infiltrated with LR White resin. Embedding is achieved by inverting gelatin capsules containing pre-cooled LR White resin (0°C), to which has been added the manufacturer's accelerator (1.5 µl per 1 ml of resin, Sect. 5.3.1.2), over the cover-slips or pieces of cover-slip. The cover-slips should be lightly drained of infiltration resin and placed cell-side up on a shiny surface such as aluminium foil. The inverted capsules on the coverslips are left at 0°C to polymerise for at least four hours. If required, further polymerisation, to increase the electron beam stability of the resin, can be achieved by leaving the capsules at 50°C for 1-2 hours. When the cover-slips, or pieces of cover-slip, are peeled gently away, the cells are left embedded in the resin. Sections of the cells will be in the plane of growth. If sections are required perpendicular to the plane of growth, the base of the LR White block containing the cells is cut off, trimmed and placed at right angles in a flat-bottomed BEEM-type capsule. The capsule is filled with fresh catalysed resin, as above, and allowed to polymerise. The new LR White block can now be shaped up and sections cut transversely through the cells. To improve their visibility and make it easier to manipulate or section them, the cells in the base of LR White blocks can be lightly stained with toluidine blue. The cells will be in the same orientation as in the monolayer, having a side which was adhered to the coverslip and a surface which was facing the culture medium.

3.1.3 Solid Tissue

The major division here is between clinical or human tissue and tissue from other sources. Human post-mortem tissue will always suffer from the effects of anoxia and autolysis, sometimes very severely, due to unavoidably prolonged delays in obtaining it. Clinical biopsy or surgically removed tissue will often be much fresher but even so considerable time may elapse before it can be dealt with. Both can only be fixed by immersion and tissue block size is critical. A brief approximation of the block sizes that the various embedding protocols can cope with is included under their descriptions. Small blocks are preferable because fixatives have to diffuse in to the tissue and large pieces will not be evenly fixed, but very small blocks give too limited a view of some tissues and can lead to sampling error. The tissue should be diced to the required size (1-3 mm^3) and placed immediately in to the chosen fixative. If the fixative is neutrally buffered aldehyde the solution will have to contain a high concentration of it (> 0.5% glutaraldehyde or 2% formaldehyde or both) and tissue should be fixed for periods of at least 2-4 hours at 22°C (room temperature) to allow it to diffuse to the centre of the tissue, which, especially when using glutaraldehyde, can be very slow (see Sect. 1.3.2.1). Following primary fixation, at least some tissue should be post-fixed in osmium tetroxide and dedicated to ultrastructure in order to provide a comparison with the the same tissue prepared for immunocytochemistry by other procedures. Artefacts due to autolysis or anoxia can then be separated from that caused by the use of alternative fixation and processing methods. (For recommendations on fixation methods see Sect. 1.3.2.1). Suggestions have been made for the resuscitation of fresh solid tissue pieces by keeping them for a short time, often a matter of days, in oxygenated culture medium. This obviously does not apply to post-mortem tissue. Some tissues recover well, some do not, meanwhile time will have been lost finding out. Even if the cells recover, changes can occur to tissue structure following culturing.

For other solid tissues more handling choices are available. Animals may be sacrificed and tissue removed rapidly by dissection for treatment by immersion fixation, as described above for biopsy material. Whole small animals can be perfused with fixative whilst very deeply anaesthetised (so deeply comatose that they have no reflex responses and cannot recover). In some cases, especially larger animals, separate organs can be isolated and perfused with fixative. Perfusion fixation does not rely on diffusion for the fixative to take effect, the fixative being delivered to tissue via its own blood This approach allows very low concentrations of aldehyde to be used so that tissue cross-linking is minimalised and its reactivity with antisera is optimised (see Sect. 1.3.2.1). Parenchymal perfusion is a method intermediate between immersion and perfusion in which fixative is dripped on to a small area of tissue, exposed by dissection, in a deeply anaesthetised animal (see above). The tissue will usually have particular problems not suited to immersion or perfusion fixation. It may be difficult to approach or have special pH or osmolarity requirements. Following drip fixation for 20-30 minutes only the small tissue area of interest is excised and fixation is continued by immersion.

3.2   Protocols Employing Full Dehydration of Tissue at Room Temperature (RT)

These methods are applicable to large blocks of tissue heavily cross-linked by glutaraldehyde (2-5 mm^3) or smaller blocks of post-osmicated tissue (1 mm^3). All the processing steps are completed at 22°C (room temperature) on a rotary device to improve tissue/solution exchange. The following suggestions would be suitable for most soft tissues to be embedded in Epon, Taab Resin or Araldite (Glauert, 1975; Hayat, 1981), or in one of the acrylic resins (LR White; Lowicryls K4M and HM20).

3.2.1 Fixation

See also Sect. 1.3.2 on Fixation Strategies. Pieces of tissue are first fixed either by immersion or vascular perfusion in a solution of 1-6% neutral-buffered (usually 0.1M phosphate or cacodylate buffer, pH 7.3) glutaraldehyde for 2-4 hours at 22°C. They are then thoroughly washed for at least 4 hours or preferably overnight in frequent changes of the buffer alone. Sucrose (2% w/v) or dextran (25,000 MW, 1.5% w/v) is frequently added to the fixative and the buffer washes (particularly for vascular perfusion) to balance with the colloid osmotic pressure of the tissue. This is probably achieved better by dextran. The tissue blocks are post-fixed in neutral-buffered (veronal acetate or cacodylate buffer, pH 7.3) 1-2% osmium tetroxide for 2-4 hours at 22°C and transfered to 70% ethanol (or acetone) to stop the osmium reaction and begin dehydration. It perhaps should be mentioned here that veronal acetate buffer has a poor buffering capacity at physiological pH.

In order to increase the reactivity of tissue with antibodies and lectins post-osmication is sometimes omitted, and the tissue fixed soley with glutaraldehyde. The subsequent tissue treatment is exactly the same except that following the buffer rinses it is placed in to 50% ethanol (or acetone) for 15 minutes before being transferred to 70% ethanol (or acetone).

Alternatively, immersion-fixation in a mixture of neutrally buffered 1% highly purified glutaraldehyde (Polysciences) and 0.2% picric acid (BGPA - Newman et al, 1982, 1983a; Newman and Jasani, 1984a) provides consistency in the preservation of membrane structure and antigenicity in large pieces of tissue, for example surgical biopsies, (Newman et al, 1982, 1983a, 1986). The tissue should be washed briefly in 50% ethanol (10 min) then placed in to two 30 minute changes of 70% ethanol to wash out picrates and excess aldehyde.

Post-fixation is also possible with 0.5-2% uranyl acetate for 1-3 hours at 22°C. After initial fixation with aldehyde and/or osmium, the tissue blocks are washed in two 15 minute changes of buffer or distilled water prior to being placed in aqueous uranyl acetate solution. Within the authors' laboratory, aqueous uranyl acetate has been found to routinely give perfectly adequate ultrastructure. However, buffered solutions (veronal acetate, [Ryter et al, 1958]; Tris-maleate, [Berryman and Rodewald, 1990]) can be used, but phosphate buffers must be avoided or unwanted precipitates will be deposited within the tissue (Sect. 8.1.6). If phosphate buffer was used in any of the earlier steps, thorough washing of the tissue blocks is recommended to remove any traces of phosphate. Following post-fixation, the tissue blocks are washed in two 15 minute changes of distilled water and are then ready for the dehydration protocols.

Heat polymerisation is inevitable when embedding in an epoxy resin and, for tissue which has been heavily fixed and/or osmium post-fixed, is the most suitable of the several alternatives when embedding in an acrylic resin. Fixation in formaldehyde, alone, or with low concentrations of glutaraldehyde and/or picric acid (Ito and Winchester, 1963) or the Periodate-Lysine-Paraformaldehyde fixative (PLP) of McLean and Nakane (1974), give, in the absence of post-fixation in osmium, results with heat polymerisation that are only suitable for light microscopy. The weak crosslinking capacity of these fixative solutions cannot stabilise tissue structure sufficiently for complete dehydration and infiltration at room temperature with polymerisation of resin by heat, and are only worth employing for electron microscopy with the less deleterious acrylic embedding methods (see protocols for either partial dehydration, Sect. 3.3, or PLT, Sect. 3.5).

3.2.1.1 Aldehyde Blocking

Following glutaraldehyde fixation, treatment of the tissue in 1% ammonium chloride after the buffer washes and before dehydration will neutralise residual aldehyde groups which could subsequently interfere with immunocytochemical reactions. However, variable amounts of tissue damage (depending on the tissue, extent and type of fixation etc.) will occur. Gentler on-section methods for dealing with residual aldehyde groups are discussed later (Sects. 8.1.4/8.1.7).

3.2.2 Protocol 1: for Epoxy Resins

All operations are carried out at 22°C on a rotary device unless otherwise stated. Each step represents a fresh change of solution.

3.2.2.1 Dehydration

A. Ethanol

1. 70% ethanol   2 x 15 min
   (the first may turn blue/black with reduced osmium).
2. 90% ethanol   15 min
3. 100% ethanol   2 x 30 min
4. Propylene Oxide   2 x 15 min
   (xylene, chloroform or trichloroethane are non-reactive solvents which may also be used; see Sect. 1.3.3).

B. Acetone
(unsuitable for BGPA fixed tissue)

1. 70% acetone   2 x 15 min
   (the first may turn blue/black with reduced osmium).
2. 90% acetone   15 min
3. 100% acetone   2 x 15 min

3.2.2.2 Infiltration

5. 1:2 neat resin*:solvent
   mixture   60 min
6. 1:1 neat resin:solvent
   mixture   60 min
7. 2:1 neat resin:solvent
   mixture   60 min
8. Neat resin*   60 min
9. Neat resin   overnight
10. Neat resin   60 min
11. Place tissue in resin-filled capsules

4. 1:1 neat resin*:acetone
   mixture   60 min
5. 2:1 neat resin:acetone
   mixture   60 min
6. Neat resin*   60 min
7. Neat resin   overnight
8. Neat resin   60 min
9. Place tissue in resin-filled capsules

3.2.2.3 Polymerisation

By heat at 60°C (Sect. 5.2.1)   By heat at 60°C (Sect. 5.2.1)

*Neat resin refers to the final mixture of epoxy resin to which has been added the relevant accelerator (Glauert, 1975, 1991; Sect. 2.1.2). The neat resin may be stored in air-tight polypropylene containers (for example 10-20 ml syringes) at -20°C for at least 1 month. N.B. Before use allow the resin to warm to room temperature to avoid condensation.

3.2.3 Protocol 2: for Acrylic Resins

This protocol can be used with LR White or Lowicryls K4M and HM20. All operations are carried out at 22°C on a rotary device unless otherwise stated. Each step represents a fresh change of solution.

3.2.3.1 Dehydration

1.	70% ethanol	2 x 15 min
	(the first may turn blue/black with reduced osmium).	
2.	90% ethanol	15 min
3.	100% ethanol	2 x 15 min

3.2.3.2 Infiltration

LR White

4.	Neat resin*	60 min
5.	Neat resin	overnight
6.	Neat resin	60 min
7.	Place tissue in resin-filled capsules	

Lowicryls

4.	1:1 neat resin**:ethanol mixture	60 min
5.	2:1 neat resin:ethanol mixture	60 min
6.	Neat resin**	60 min
7.	Neat resin	overnight
8.	Neat resin	120 min
9.	Place tissue in resin-filled capsules	

*The hard-grade (or original, as supplied by the manufacturer complete with added catalyst; Sect. 2.2.1.3) resin is recommended for cytochemistry and immunocytochemistry so as to preserve good sectioning qualities while minimising the extent of resin cross-linking during polymerisation (Chap. 5).
**The resin is prepared complete (i.e. crosslinker plus monomer; see Varying the Hardness of the Resins in Sect. 2.2.3.3) with the relevant amount of benzoyl peroxide (Sect. 5.1).

3.2.3.3 Polymerisation

For heavily fixed fully dehydrated tissue heat polymerisation is strongly recommended. However, it can also be by any of the other methods involving chemical catalytic or UV light procedures found in Chap. 5:-

(i)   Heat - LR White (Sect. 5.2.2.1); Lowicryls (Sect. 5.2.2.2).
(ii)   Chemical catalytic methods at room temperature - LR White (Sect. 5.3.1.1); Lowicryls (Sect. 5.3.2.1).
(iii)   UV light at room temperature - (<u>not</u> for osmium fixed or heavily pigmented tissue) see Sect. 5.4.2.1 for LR White and Lowicryls.

3.3   Protocols Employing Partial Dehydration of Tissue at Room Temperature (RT)

3.3.1 Fixation

The levels of antigenic sensitivity that can be retained within tissue often depend upon the duration of fixation, the concentration of fixative chosen and the method of its application (see Sect. 1.3.2). *The subsequent method of dehydration, infiltration and polymerisation should, therefore, be tailored to these initial choices.* Partial dehydration is most profitably employed with tissue that has not been totally cross-linked by aldehydes, and it is now realised that heat polymerisation, which is very extractive, is best avoided (Sect. 1.3.1). Partial dehydration is not suitable for epoxy resins (Sect. 1.3.3).

Protocols employing partial dehydration and chemical catalytic polymerisation of acrylic resins can be performed at room temperature (22°C) or in the cold from 0°C down to -35°C or lower depending on the freezing point of the resin. Room temperature protocols employing chemical catalytic polymerisation are rapid techniques that can be completed on the same day (e.g in half a working day; omitting the time taken for fixation, the blocks are ready for sectioning after 2-2.5 hours). However, they are more extractive and not as sensitive in retaining antigenic immunoreactivity as the lower temperature processing protocols and polymerisation methods. Room temperature chemical catalytic polymerisation protocols are, therefore, preferred for tissue which has been more heavily crosslinked, with higher concentrations of fixative (Sect. 1.3.2). Methods at 0°C are slower taking over 6 hours, leaving out the time taken for fixation, and, as the processing and polymerisation temperature is reduced so the block size also has to be reduced and the method becomes more extended in time. However, low temperature protocols are able to take full advantage of the benefits of employing low fixative concentrations (Sect. 1.3.2.1). Generally, partial dehydration and chemical catalytic polymerisation have only been used extensively at room temperature and 0°C. For tissue processing below -20°C, PLT and either blue- or, more usually, UV-light polymerisation have been preferred. However, it is possible to carry out partial dehydration below 0°C and down to -35°C, but extreme caution must be used to ensure that freezing out of water does not occur due to its increased concentration in 2:1 neat resin/70% organic solvent mixtures (J.A. Hobot, unpublished results). The advantages, if any, of using this variation of partial dehydration remain to be elucidated.

As partial dehydration stops at a concentration of 70% organic solvent, this procedure is *only suitable* for the *more polar acrylic resins* able to tolerate at least up to 10% water (i.e. **LR White, Lowicryls K4M and K11M**). Initially it was only used in conjunction with LR White (Newman et al, 1982, 1983a; Newman and Jasani, 1984b; Yoshimura et al, 1986), but has now also been introduced for Lowicryl K4M (Hobot, 1990; Hobot and Newman, 1991). This again demonstrates how the versatility of acrylic resins allows the experimenter to devise his own approaches for specialised requirements. Lowicryl K4M will tolerate even greater amounts of residual water in incompletely dehydrated tissue than LR White.

3.3.1.1 Fixation for Room Temperature Protocols

These protocols are applicable to intermediate-sized blocks of aldehyde fixed tissue (1-2 mm^3), and small blocks (<1 mm^3) of post-osmicated tissue. For fixation, tissue should be completely cross-linked in glutaraldehyde or <u>B</u>uffered <u>G</u>lutaraldehyde <u>P</u>icric <u>A</u>cid (BGPA) (2-4 hours) by either vascular perfusion or immersion methods at 22°C. If formaldehyde, formaldehyde/glutaraldehyde/(/picric acid) mixtures or periodate/lysine/paraformaldehyde (PLP) are used, fix for at least 24 hours. Very rapid fixation (5-10 minutes) is possible by microwave stimulation (Hopwood et al, 1984; Jones and Gwynn, 1991). If picric acid-fixed tissue is washed in buffer (or any aqueous solution) insoluble picrates are deposited so it must be placed directly in to 50% ethanol to dissolve out the picrates and wash away unbound glutaraldehyde. Tissue fixed in glutaraldehyde alone is washed thoroughly in several changes of buffer for at least 4 hours or overnight. Tissue fixed in formaldehyde should not be given such extensive buffer washing (overnight) or reversal of fixation may occur (Tissue Fixation in Sect. 1.3.2.1). Fixation reversal can lead to improved antibody response but it can also be responsible for severe antigen extraction and loss of tissue structure. After buffer washes, and possible abolition of aldehyde groups (Sect. 3.2.1.1), aldehyde fixed tissue is placed in to 50% ethanol. Post-osmicated tissue is placed directly in to 70% ethanol to stop fixation.

Post-fixation is also possible with 0.5-2% uranyl acetate for 1-3 hours at 22°C as detailed above in Sect. 3.2.1.

3.3.1.2 Fixation for Cold (0°C) Temperature Protocols

The cold (0°C) temperature protocol is applicable to small tissue blocks (<1mm^3) and is unsuitable for post-osmicated tissue. This method is an adaptation of protocol 1 (Sect. 3.3.2), and was developed for tissue which has been lightly fixed in order to preserve its reactivity with antibodies and lectins (Newman and Hobot, 1987; Newman, 1989). Unfortunately, such tissue also often retains active groups which affect the chemical polymerisation of the resin. If they are not completely cross-linked by fixation, naturally occurring accelerators will cause the resin to begin gelling in the tissue during infiltration with resin, particularly if warmed. When coupled with the accelerator added for polymerisation, an exaggerated response causes the outside of the block to polymerise so rapidly that it becomes shrunken and impenetrable whilst the centre stays soft and swells. When it occurs, this effect is clearly detectable in toluidine blue-stained semithin sections. It is overcome by infiltrating the tissue with resin only at room temperature, and carrying out the polymerisation at 0°C. This allows a longer time for the accelerator to infiltrate the tissue before the onset of gelation. At 0°C, the resin begins to gel after 2-3 hours prior to hardening (Newman and Hobot, 1987).

Fixation is by vascular perfusion (preferable) or immersion (of small/thin specimens) at 22°C using low concentrations of fixative (e.g. 0.1-0.2% glutaraldehyde) for 15-60 min. Tissue that has been delicately fixed is liable to extraction and distortion and is, therefore, required to be treated with more care than the heavily cross-linked tissue whose embedding is described by the previous protocols. Short buffer washes (four 15 minute changes) are recommended. Tissue fixed in neutral-buffered formaldehyde alone for short periods (i.e. < 4 hours) should be washed in several changes of buffer for 1-2 hours. If a small percentage of glutaraldehyde was also employed, a longer buffer wash of 3 hours is preferable but even in this case prolonged washing (overnight) can lead to fixation reversal (Tissue Fixation in Sect. 1.3.2.1). It should be remembered that from present evidence the level of antigenicity that can be retained within tissue by even supposedly "mild" or low concentrations of glutaraldehyde and formaldehyde mixtures is often the same as that obtained after 15-60 min fixation with 1% glutaraldehyde (Sect. 1.3.2).

Post-fixation is also possible with 0.5-2% uranyl acetate for 1-3 hours at 22°C. After initial fixation with aldehyde, the tissue blocks are washed in two 15 minute changes of buffer or distilled water prior to being placed in aqueous uranyl acetate solution and processed further as detailed above in Sect. 3.2.1.

3.3.2 Protocol 1: Room Temperature Rapid Polymerisation

Room temperature chemical catalytic polymerisation methods are very rapid embedding procedures (2-2.5 hours; Yoshimura et al, 1986; Newman and Hobot, 1987). These procedures have been devised to avoid the prolonged exposure of tissue to the highly extractive monomeric resin which is a feature of the heat cure method (Sect. 1.3.1). When compared with sections of tissue prepared by heat polymerisation, much lower concentrations of antisera can often be used to achieve the same degree of labelling, indicating a rise in sensitivity.

Unless otherwise stated, all the following steps are completed at 22°C (room temperature) on a rotary device to improve tissue/solution exchange.

3.3.2.1 Dehydration

Tissue fixed in:

	Aldehyde	Osmium	BGPA
Buffer washes	4h to overnight	-	-
1. 50% Ethanol	1 x 10 min	-	1 x 10 min
2. 70% Ethanol	2 x 15 min	3 x 10 min	2 x 30 min

3.3.2.2 Infiltration

3. Neat resin*   4 x 20 min (37°C optional)

*When using LR White, the hard-grade (or original) resin is recommended for cytochemistry and immunocytochemistry so as to preserve good sectioning qualities while minimising the extent of resin cross-linking during polymerisation (Chap. 5). When using Lowicryl K4M, the resin is prepared complete (i.e. crosslinker plus monomer; see Varying the Hardness of the Resins in Sect. 2.2.3.3) with the relevant amount of benzoyl peroxide (Sect. 5.1)

The resins may be prewarmed to 37°C if desired to speed infiltration of large blocks or dense tissue.

3.3.2.3 Polymerisation

Polymerisation at room temperature by the chemical catalytic method (Sect. 5.3) takes only about 20-30 minutes for LR White (Sect. 5.3.1.1) or for Lowicryl K4M (Sect. 5.3.2.1). UV light at room temperature can also be used but is much slower (Sect. 5.4.2.1).

3.3.3 Protocol 2: Cold (0°C) Polymerisation

All the following steps are completed on a rotary device at 22°C (room temperature) unless otherwise stated.

3.3.3.1 Dehydration

1.	50% ethanol	1 x 10 min
2.	70% ethanol	3 x 10 min

3.3.3.2 Infiltration

3.	2:1 neat resin**:70% ethanol*	1 x 30 min
4.	Neat resin (LR White, K4M)**	4 x 20 min

*An intermediate dilution of resin in 70% ethanol is recommended to (a) speed infiltration, and (b) reduce osmotic shock and uneven tissue shrinkage. To prepare a clear solution, with either resin, the 70% ethanol must be made from 100% (anhydrous) ethanol. When using LR White, old resin stocks must not be used. The LR White (hard-grade) must have been recently purchased (< than 3 months old and stored correctly at 4°C) and allowed to reach 22°C (room temperature) before mixing. A slight turbidity (in the case of LR White) is unimportant and will not interfere with infiltration. A 3:1 mixture of LR White:70% ethanol, although not quite as effective, always remains clear.

**When using LR White, the hard-grade (or original) resin is recommended for cytochemistry and immunocytochemistry so as to preserve good sectioning qualities while minimising the extent of resin cross-linking during polymerisation (Chap. 5). When using Lowicryl K4M, the resin is prepared complete (i.e. crosslinker plus monmer; see Varying the Hardness of the Resins in Sect. 2.2.3.3) with the relevant amount of benzoyl peroxide (Sect. 5.1)

3.3.3.3 Polymerisation

Polymerisation at 0°C is by the chemical catalytic method (Sect. 5.3) and takes about 3-4 hours for either LR White (Sect. 5.3.1.2) or Lowicryl K4M (Sect. 5.3.2.2). It is recommended that the blocks are left overnight at 0°C and stored at -20°C or lower (Sect. 6.2.2). Further polymerisation may be necessary to produce beam stability of sections (Sects. 5.3.1.2/5.3.2.2). UV light can also be used at 0°C but is much slower (Sect. 5.4.2.2).

3.4 Protocols Employing Dehydration of Tissue at Cold Temperatures (0°C to -20°C)

Low temperature procedures are designed to reduce the damage to tissue caused by dehydration and embedding so that cross-linking with aldehyde fixatives can be reduced to a minimum to preserve tissue reactivity (Sects. 1.3.3/3.3.1). Some doubt exists, however, as to whether -20°C is low enough to significantly reduce tissue losses in very lightly fixed tissue. Probably the degree of fixation required to preserve reasonable levels of ultrastructure in most tissue would be on a par with that used for partial dehydration processing and embedding protocols. These protocols can be employed with **LR White, LR Gold** or **Lowicryls.**

3.4.1 Fixation

These protocols are applicable to very small blocks of unpigmented tissue (<0.5 mm^3). They are not suitable for osmicated tissue. Tissue is fixed either by vascular perfusion or immersion with very low concentrations of fixative (0.1-0.2% glutaraldehyde) for 15-60 min. Glutaraldehyde/formaldehyde mixtures can be used and buffer washes are kept to a minimum (see Sect. 3.3.1.2)

3.4.2 Protocol for Dehydration down to -20°C

3.4.2.1 Dehydration*

1.	30%	ethanol	0°C	30 min
2.	50%	ethanol	-20°C	60 min
3.	70%	ethanol	-20°C	60 min
4.	100%	ethanol	-20°C	60 min x 2

*All the dehydration steps may also be carried out at either 0°C or -10°C

3.4.2.2 Infiltration*

5.	1:1 neat resin**:ethanol	-20°C	60 min
6.	2:1 neat resin:ethanol	-20°C	60 min
7.	Neat resin**	-20°C	60 min
8.	Neat resin	-20°C	overnight
9.	Neat resin	-20°C	for 60 min

*All the infiltration steps may also be carried out at either 0°C or -10°C
**When using LR White, the hard-grade (or original) resin is recommended for cytochemistry and immunocytochemistry so as to preserve good sectioning qualities while minimising the extent of resin cross-linking during polymerisation (Chap. 5). Lowicryls K4M or HM20 complete with crosslinker and monomer (see Varying the Hardness of the Resins in Sect. 2.2.3.3).

3.4.2.3 Polymerisation

Polymerisation at 0°C, -10° or -20°C is by UV light (Sect. 5.4.2.2) or by chemical catalytic methods (see Sect. 5.3 on Chemical Catalytic Methods including sections on Cold Temperatures [-20°C] for guidance on how to proceed empirically for L.R.

White [Sect. 5.3.1.3] and Lowicryls [Sect. 5.3.2.3]). For blue-light polymerisation of LR Gold see Sect. 2.2.2.

3.5 Protocols Employing Dehydration of Tissue at Progressively Lower Temperatures (PLT: -35°C to -50°C)

These protocols for the Progressive Lowering of Temperature technique (PLT) are only suitable for the **Lowicryl resins**.

3.5.1 Apparatus for PLT

Two alternatives exist for carrying out the steps below 0°C:-

(i)   the purchase of equipment that will reach and work over the range of temperatures required (0°C to -50°C);

(ii)   the use of cold mixtures.

A combination of the two is also possible.

Commercially available pieces of equipment are manufactured by Balzers Union (LTE 020 Low Temperature Embedding Apparatus) and Leica (Reichert CS Auto). The latter was initially manufactured for cryosubstitution procedures, but has recently been introduced for PLT as well. (A modified version of the CS Auto is now available just for PLT). Its advantage is that besides being able to carry out the dehydration and resin infiltration steps at low temperatures, it can also perform UV light polymerisation of the resin blocks as well. A disadvantage is that the specimens can only be processed with one organic solvent and one resin of choice. This is not the case for the equipment produced by Balzers Union. Here there are four separate wells for specimen vials, with two spare wells for precooling the solutions. Thus up to four different organic solvents or resins can be used if desired. However, no capacity for UV polymerisation exists. In both pieces of equipment there is provision for agitation/stirring of the infiltrating solutions to facilitate solvent/resin exchange.

Alternatively, refrigerators or freezers capable of being set to one of the required temperatures needed for PLT (0°C, -20°C, -35°C, -50°C) can be used. Chest-type deep-freezes are more suitable as they warm up much more slowly than upright appliances. Gentle hand agitation of the specimen vials, performed at regular intervals, aids infiltration/exchange of solvent solutions. UV lights can be mounted either inside or on top of the freezers (Sect. 5.4.1).

Cold mixtures will also allow the necessary low temperatures to be reached. Ice will give 0°C, and an ice/salt (sodium chloride) mixture of ratio 3:1 (w/w) gives -20°C. For -35°C, a mixture of crushed ice and crushed calcium chloride ($CaCl_2.6H_2O$) in a ratio of 1:1.44 (w/w) is used. The mixtures of o-xylene and m-xylene, suggested in the pamphlet accompaning the Lowicryl kits, and which can reach temperatures of -35°C to -50°C, are not recommended because of their toxic vapours. The specimen vials can be placed in an aluminium block which has suitable holes drilled in it (Sect. 4.3.3). The aluminium block can then be placed safely in the appropriate cold mixture.

3.5.2 Fixation

Sub -30°C temperatures provide greater protection against tissue extraction and protein conformational change during embedding, so that aldehyde crosslinking can be reduced correspondingly. As a consequence, the tissue may retain high levels of reactivity with antisera and lectins (Sect. 1.3.3). Unless there is a special reason the use of glutaraldehyde solutions above 0.2% for longer than 60 min should be avoided (Sect. 1.3.2). More heavily cross-linked tissue can be dealt with just as successfully by other simpler methods (Newman and Hobot, 1989; Hobot and Newman, 1991; Sects. 3.2.2/3.2.3).

PLT is applicable to very small blocks of unpigmented tissue (<1 mm^3) and is not suitable for osmicated tissue (see Sect. 3.5.3.3). Tissue for PLT to -30°C/-50°C (Sects. 3.5.3/3.5.4) is fixed by vascular perfusion or immersion with glutaraldehyde concentrations of 0.1-0.2% for 15-60 min (see Sect. 3.3.1.2) followed by short buffer washes (two to four for 15 min).

Post-fixation is also possible with 0.5-2% uranyl acetate for 1-3 hours at 22°C. After initial fixation with aldehyde, the tissue blocks are washed in two 15 minute changes of buffer or distilled water prior to being placed in aqueous uranyl acetate solution and processed further as detailed in Sect. 3.2.1.

3.5.3 Protocol 1: PLT to -35°C

3.5.3.1 Dehydration*

1.	30%	ethanol	0°C	30 min
2.	50%	ethanol	-20°C	60 min
3.	70%	ethanol	-35°C	60 min
4.	100%	ethanol	-35°C	60 min x 2

*Like other acrylic resins the Lowicryls can tolerate small residues of water in tissue so that it is not essential to fully dehydrate beyond 90% solvent (Sect. 1.3.3).

3.5.3.2 Infiltration

5.	1:1 neat resin*:ethanol	-35°C	60 min
6.	2:1 neat resin:ethanol	-35°C	60 min
7.	Neat resin*	-35°C	60 min
8.	Neat resin	-35°C	overnight
9.	Neat resin	-35°C	for 60-120 min

*Lowicryls K4M or HM20 complete with crosslinker and monomer (see Varying the Hardness of the Resins in Sect. 2.2.3.3).

3.5.3.3 Polymerisation

The resins are polymerised by UV-light at -35°C/-50°C (Sect. 5.4.2.3). Blocks are sectionable after the first 24 hour period of illumination, but are improved by further polymerisation.

Some success in embedding very small pieces of osmicated tissue in Lowicryl HM20 at low temperature using UV light polymerisation has been recorded following PLT (Hobot et al, 1982) or cryosubstitution protocols (Humbel et al, 1983). However, specimens prepared by PLT, but post-fixed in osmium or appearing pigmented and unsuitable for UV polymerisation, can be chemically polymerised at low temperatures (Sect. 5.3.2.4) although this method is not recommended for general use.

3.5.4 Protocol 2: PLT to -50°C

The embedding of tissue at -50°C is only possible with HM20.

3.5.4.1 Dehydration

1.	30%	ethanol	0°C	30 min
2.	50%	ethanol	-20°C	60 min
3.	70%	ethanol	-35°C	60 min
4.	90%	ethanol	-50°C	60 min
5.	100%	ethanol	-50°C	60 min x 2

3.5.4.2 Infiltration

7.	1:1 neat resin*:ethanol	-50°C	60 min
8.	2:1 neat resin:ethanol	-50°C	60 min

9.	Neat resin*	-50°C	60 min
10.	Neat resin	-50°C	overnight
11.	Neat resin	-50°C	for 60-120 min

*Lowicryl HM20 complete with crosslinker and monomer (see Varying the Hardness of the Resins in Sect. 2.2.3.3).

3.5.4.3 Polymerisation

Lowicryl HM20 is polymerised at -50°C by UV-light (Sect. 5.4.2.3; see also Sect. 3.5.3.3).

4   Cryotechniques

Cryotechniques do not require tissue to be stabilised by fixatives prior to resin embedding, and so are useful for preserving antigenic sites that are sensitive to even low concentrations of fixative, or for preserving structures that cannot be stabilised during fixation or dehydration (Sect. 1.3.2). The prerequisite for cryotechniques is for biological tissue to be rapidly frozen prior to any processing steps. Processing can then be in to resin (cryosubstitution, Sect. 4.3; freeze-drying, Sect. 4.6), or the frozen tissue can be sectioned directly (cryoultramicrotomy, Sect. 4.9).

4.1 Tissue Preparation for Freezing

Biological tissue that is to be frozen is not pretreated with any chemical fixatives, as these not only cross-link cell components, but also cause a breakdown in membrane permeability barriers so inducing ionic leakage and a possible subsequent reorganisation of cellular structures (Sect. 1.3.2.2; Hobot et al, 1985; Woldringh, 1973). Cell-fractions, cell suspensions, cell cultures and monolayers, solid tissue and many other kinds of tissue preparation are acceptable for freezing. Tissue pieces suitable for this method have to be very small indeed (no more than 0.1-0.2 mm in size) and even then only 10-12 μm of the block may be adequately preserved. High-pressure freezing is capable of dealing with larger pieces of tissue and can preserve up to 200-500 μm without visible ice-crystal damage (Studer et al, 1989), although it would be preferable to keep sample size to a minimum.

It cannot be stressed enough that the single, most important step in freezing biological specimens is the initial preparation of the sample and its transfer to the freezing apparatus. Growing cell cultures (e.g. bacteria, yeasts, viruses), cell suspensions, isolated cell components, botanical or fungal specimens, are relatively easily handled (Sect. 3.1.1). Growing cell suspensions can be harvested by filtration and aeration, and can be frozen directly when still on the filter. If the filter does not embed well in the resin system chosen, they can be transferred by smearing on to thinner paper (e.g. cigarette paper; Hobot et al, 1985, 1987). Filtration is an improvement on centrifugation for harvesting growing cells or even cell suspensions, because the anaerobic conditions induced by centrifugation, and leading to ionic leakage, are prevented (Sect. 3.1.1). Rapid freezing is carried out immediately after filtration, with any excess water or fluid removed from the specimen by gently touching the sample's surface with filter paper. Care must not only be taken to prevent drying out the specimen during this procedure, but also to avoid air-drying of the sample while mounting it on to the freezing apparatus's specimen holder. Air-drying will cause damage to ultrastructure. Also, drying/evapouration can lead to an increase in solute concentration in the medium surrounding the specimen, causing changes in the external osmotic pressure which will result in cellular damage or reorganisation. Resuspending cell suspensions or pellets in matrices such as gelatin, agarose or agar is an attractive alternative, but it should be borne in mind that these gels are highly hydrated compounds, containing up to 98% or more water, and this could result in extracellular ice crystals forming and damaging the tissue. Judging the "correct" amount of moisture to leave on a specimen immediately prior to freezing is best handled empirically, with perhaps a few trial runs being necessary to find the right conditions for freezing. It is also good working practice to have all the appropriate pieces of equipment and solutions required ready prior to commencing any freezing experiment.

Tissue cell cultures or monolayers (Sect. 3.1.2) can be carefully excised, together with the support film they are growing on, from their culture dishes and mounted on to the freezing appparatus's specimen holder. Better still is to have the cell cultures or monolayers already growing on pieces of support film (e.g. Thermanox) cut to the correct dimensions required for freezing. These can then be handled more easily, quickly and conveniently for rapid freezing. Any excess fluid or culture medium must be removed as described above, but bearing in mind from the preceding discussion the associated problems that can arise if not carried out correctly. A way to overcome these problems is to employ high-pressure freezing, where the removal of all of the culture medium is not necessary, and thus causes the samples no adverse disturbance (Sect. 4.2.5).

Greater problems are be faced when freezing complex organ tissue (e.g. pancreas, liver, kidney etc.), as alterations to cellular organisation due to autolysis or oxygen starvation occur rapidly. No matter how sophisticated all the following procedures are, they will not overturn the damage that can be done at the outset. A great deal of planning on how best to handle the biological specimen under study, in order to cause it as little damage or change as possible while bringing it to the freezing apparatus, should precede the use of cryotechniques. Tissue samples have to be quickly dissected out of the killed or deeply anaesthetised animal (so deeply comatose that they have no reflex responses and cannot recover), cut in to very small pieces and frozen rapidly. Excess water, tissue fluid or buffer solution must not be left surrounding the specimen. This will seriously affect the quality of freezing, causing ice crystal damage. Such fluid has to be carefully removed with fibre-free filter paper, but not so as to completely dry the specimen and induce drying artefacts, or increase the concentration of solute in the surrounding medium and cause osmotic changes in the tissue and cells. This level of control is acquired by experience in handling various specimens, and so may necessitate several freezing runs to optimise the conditions required for any given sample. As the tissue is cut mechanically in to the correct size and shape for freezing, the surface of the tissue is somewhat disfigured during the cutting process. It is also this surface that yields well-frozen material (Sect. 4.2.5), and therefore a reduction in the amount of well-preserved ultrastructure for examination will be the result.

It must be stressed that a tremendously limiting factor is the time taken from the moment of death or of anaesthesia to dissect out and prepare the tiny pieces of tissue required for freezing. Changes in cellular structures can occur very quickly due to anoxia and autolysis, such that the overall organisation of the cells following either cryosubstitution (high pressure freezing, Studer et al, 1989; impact freezing, Hobot and Newman, 1991) or freeze-drying (impact freezing, Roos et al, 1990) may be no better than that obtained after immersion fixation with glutaraldehyde. The use of precooled, hand-held "cold pliers" applied to the dead or comatose animal apparently does not improve tissue preservation (Roos et al, 1990). An automated version (Cryosnapper, Gatan Model 669, RMC, Tucson, USA) has not yet demonstrated any improvement.

The use of cryoprotectants is not advisable. The main reason for using cryotechniques is to avoid the adverse affects of all chemicals not just chemical fixatives, and so freezing apparatuses capable of reaching very high cooling rates are employed to avoid the formation of damaging ice crystals in fresh tissue. Cryoprotectants may cause structural damage to tissue, or have an affect on biological samples similar to that of fixatives (Introduction under Sect. 1.3.2.2), and, therefore, should be avoided.

4.2   Rapid Freezing and Apparatus Requirements

Cryosubstitution and freeze-drying rely on very rapid freezing methods with high cooling rates (Bald, 1985; Escaig, 1982), which have been shown to produce non-crystalline or amorphous ice (Dubochet et al, 1982). This prevents ice-crystal damage and immobilises cell substances almost instantly. There are basically five main ways of achieving this state: plunge-freezing, propane jet freezing, spray freezing, impact freezing and high-pressure freezing (for detailed reviews see: Menco, 1986; Plattner and Bachmann, 1982; Robards and Sleytr, 1985; Sitte et al, 1987a).

4.2.1 Plunge-freezing

The specimen to be frozen is directed in to liquid propane or ethane (Escaig, 1982; Robards and Sleytr, 1985). For maximum efficiency of freezing, and to reduce the time spent travelling through the cold gas layers above the coolant, the rate of specimen entry is often physically accelerated. The depth that the specimen travels in to the coolant is also important. The method requires a large reservoir (minimum depth of about 10 cm) of liquid propane or ethane in to which the specimen should be deeply plunged (at least half-way in to the coolant). Stirring or agitation of the coolant lowers the temperature slightly and improves the contact between "fresh" coolant and specimen.

When not using thin films deposited on bare grids, the specimens should be mounted on backing material that does not hinder rapid cooling rates, e.g. mica. This will also prevent curling of the specimen upon plunging, if it is harvested or placed on filter paper etc. The material to be frozen is held in place by forceps possessing long, thin tips in order to minimise thermal contact between the specimen and holder, and to ensure the attainment of a reasonable plunge depth in to the coolant. Alternatively, the specimen, especially solid tissue, can be mounted on small aluminium pins. Detailed drawings of various posibilities for holding the specimen are to be found in Sitte et al (1986, 1987a), and an in-depth review of plunge-freezing is covered by Ryan (1992).

4.2.2 Propane Jet Freezing

The specimen is mounted between two very thin copper plates, as in a sandwich, and liquid propane is squirted as a jet on to either one or both outer surfaces of the copper sandwich. This method can also be very useful for complementary freeze-fracture techniques, as after freezing the two copper plates are separated revealing two inner frozen surfaces for replication, and either one or two well-frozen surfaces (those next to the copper plates) for resin processing (Knoll et al, 1982; Müller et al, 1980; Plattner and Knoll, 1984).

4.2.3 Spray Freezing

The specimen to be frozen, invariably as a suspension in a unicellular or particulate form, is rapidly sprayed via a fine nozzle in to liquid propane (Bachmann and Schmitt, 1971). The temperature of the liqiud propane is raised to -85°C and it is allowed to evaporate away under vacuum. The frozen droplets can be recovered and processed further.

4.2.4 Slam or Impact Freezing

The specimen is brought rapidly in to contact with a cold metal surface (metal mirror freezing). The metal is generally pure copper, usually cooled with liquid nitrogen or, less commonly, liquid helium (Escaig, 1982; Heuser et al, 1979; Van Harreveld and Crowell, 1964). The specimen, mounted on a mica backing placed on the apparatus's plunger, is brought down very rapidly on to the cold metal surface. At contact, the cooling rate is at present far faster than that achieved by the other methods discussed here (Bald, 1983; Escaig, 1982), although continuing improvements to technology and specimen preparation may change in favour of, for example, jet freezing (Knoll et al, 1982; Plattner and Knoll, 1984).

4.2.5 High-Pressure Freezing

The specimen to be frozen is placed in a small chamber and immediately prior to freezing by pressurised liquid nitrogen is subjected to a pressure of 2500 atmospheres (Moor, 1987; Studer et al, 1989). However, the equipment is expensive and freezing occurs at a much slower rate than with other methods which makes it less suitable for studying dynamic changes in cells.

The advantage in using high-pressure freezing is that the depth of well-frozen tissue which is relatively free of ice crystal damage, so providing good ultrastructure, can be up to 200 μm, and in a few favourable cases up to 500 μm. This is considerably more than the values obtained with the other methods, as for example with plunge-freezing, 5-6 μm; with impact freezing, 10-12 μm; or with jet freezing, 10-12 μm on each surface of the sandwich. Higher values have been registered, but generally for specimens which have been cryoprotected or are most favourable in their composition and size for freezing studies.

4.2.6 Source of Apparatus

A competent laboratory workshop can manufacture relatively inexpensive and efficient freezing apparatus, especially of the plunge or impact variety (see Menco, 1986; Plattner and Bachmann, 1982; Robards and Sleytr, 1985; Sitte et al, 1987a). There are several commercially available models that can be purchased after trials, produced by, for example, Balzers (Liechtenstein), Reichert-Jung (Leica, Milton Keynes, U.K.) and RMC (Tuscon, Arizona). It is most important to freeze reliably and reproducibility and the purchase of an appropriate freezing apparatus should take precedence in costings over secondary, ancillary equipment for cryosubstitution (Sect. 4.3.3).

4.2.7 Safety

A good safety code is an absolute requirement when working with cryotechniques, especially where cold liquids are being handled. Protective clothing including gloves and eye-shields should be worn. The disposal of cold liquids can be hazardous. For example, liqiud propane, when left over from the plunge method has to be disposed of carefully, as propane gas is highly explosive. It is heavier than air and sinks to the floor where it could accumulate, especially in enclosed basement premises. In well-ventilated rooms it disperses quickly, but any reservoir of liquid propane should not be heated or warmed, but allowed to come to room temperature in a well-ventilated spot away from any possible sources of combustion/ignition. Storage of propane cylinders is, therefore, of great concern for safety reasons. The Leica (Reichert-Jung) KF 80 plunge freezer comes complete with a safety burner for disposal of excess propane gas.

Liquid nitrogen used in a confined, badly-ventilated room can also be dangerous, as gaseous nitrogen can build up reducing the relative proportion of oxygen, and asphyxiation can occur rapidly, without warning. This should be borne in mind not only for the above freezing methods, but also for the use of liquid nitrogen in cryoultramicrotomy techniques (Sect. 4.9.2).

Information about safety for cryotechniques can also be found in Robards and Sleytr (1985), Ryan and Liddicoat (1987), and Sitte et al (1987b).

4.3   Cryosubstitution

There is a wide variety of protocols to choose from in the literature, but any choice should be based upon an understanding of the steps involved.

4.3.1 The Substitution Medium

Tissue that has been very rapidly frozen is stored in liquid nitrogen taking care not to allow water vapour to collect and condense on it during manipulations. The specimens are quickly transferred to the cold substitution medium (-80°C to -90°C) using forceps precooled to liquid nitrogen temperature.

The substitution medium is usually pure acetone (freezing point -95°C) or pure methanol (freezing point -94°C). However, for maximum lipid retention and to reduce the possibility of extraction of soluble tissue components it is preferable to use acetone (Weibull et al, 1984; Weibull and Christiansson, 1986; Sect. 1.3.3). Acetone cannot remove ice if contaminated with even small amounts of water (<1%) except in the presence of anhydrous molecular sieve (0.4 nm pore size, perlform or small pellets, added when the acetone medium is at -35°C) when it gives good results (Dahmen and Hobot, 1986; Hobot et al, 1985, 1987). Methanol is effective with or without molecular sieve (0.3 nm pore size) even in the presence of 10% water (Humbel et al, 1983). The molecular sieve is added to the cooled (-30°C) substitution medium.

At present, many published protocols still employ the addition of fixatives to the substitution medium. In particular it is important to omit osmium from the substitution medium when embedding in Lowicryl resins because it can hinder UV-light penetration in to tissue and mar polymerisation. Chemical polymerisation, though, can be used as an alternative (Sect. 5.3). However, osmium can also mask antigenic sites, and, therefore, reduce the level of immunoreactivity retained within the tissue (Sect. 1.2). For on-section immunocytochemistry it would have to be removed (Sect. 8.1.5). Glutaraldehyde, however, can be added and may stabilise tissue structure when substitution is in its final stage in the temperature range -50°C to -30°C (Humbel and Muller, 1984; Horowitz et al, 1990) often without adversely affecting immunoantigenicity (Carlemalm et al, 1985b; Hobot et al, 1987; Hunziker and Herrmann, 1987). A 3% solution (v/v) can be made, prior to cooling to the substitution temperature, by adding 0.6 ml of 50% vacuum distilled, highly purified glutaraldehyde (Polysciences) to 9.4 ml of organic solvent. In preparing substitution media with osmium tetroxide (0.1-2.5%) and/or uranyl acetate (0.1-0.5%), the two compouds are added separately to the organic solvent, allowed to cool to -30°C, and only then mixed together and cooled further to -80°C/-90°C. The addition of uranyl acetate to methanol greatly improve the lipid retaining qualities of the medium (Humbel, 1984; Sect. 2.2.3.2).

Schwarz and Humbel (1989) have shown that omitting fixatives from the substitution medium is important in that it can lead to an increase in labelling response. It should become routine to omit the presence of chemical fixatives and complete the final embedding in one of the Lowicryls at below -30°C (discussed also under Sect. 1.3.2.2). Prolonging the time of substitution in acetone (see Sect. 4.3.2 below) may help to improve structural preservation. As discussed in Hobot and Newman (1991), acetone reacts with -NH$_2$ groups in much the same way as aldehydes. Therefore some fixation of tissue can occur, resulting in an increase in tissue stabilisation, an effect not experienced with methanol.

4.3.2 The Temperature and Duration of Substitution

The sustitution is carried out in the medium of choice at -80°C (to -90°C) for 72-96 hours, or longer for reasons discussed above (Sect. 4.3.1). It is important to ensure that all the ice has been removed (substituted), for at temperatures from approximately -70°C upwards ice recrystallisation can occur in badly substituted samples. Therefore long substitution times are favoured, although short times may well be satisfactory depending upon sample size, and the ability of the substituting medium to remove ice and absorb water (Sect. 4.3.1). For embedding in Lowicryls K4M and HM20, the tissue is warmed up to -35°C over a period of at least 1 hour and kept at this temperature for a further 2 hours. The time over which the temperature is raised from -80°C to -35°C may not be so critical if the substitution time at -80°C is adequate (3-7 days). Steinbrecht (1982) has made a careful study of the parameters discussed in this section.

4.3.3 Apparatus for Substitution

Cryosubstitution equipment (CS-Auto, Leica [Reichert-Jung]; FSU 010, Balzers Union; MS 6000, RMC) can be purchased to achieve the very low temperatures required for substitution. They have automatic temperature control so that the duration and extent of temperature rise or fall can be reproduced reliably. Low temperatures down to -90°C can also be achieved by certain makes of commercial low temperature cabinets (Merck; Fisons Scientific Equipment).

Alternatively, -80°C can be achieved manually with a mixture of acetone and crushed or pelleted dry-ice contained in a large (7 litre) Dewar with a sealable top. Polypropylene conical centrifuge tubes with a 50 ml capacity (J. Bibby Ltd.; catalogue no. 25330B with caps) containing 10-15 ml of the substitution medium are precooled prior to use, after which the frozen tissue is added. A large, aluminium block drilled with appropriately sized holes is useful to accommodate the tubes in the dry-ice/acetone mixture (Sect. 3.5.1). The whole assembly will maintain its temperature for longer if it is contained in a deep-freeze set at -35°C. The Dewar should be topped up with dry-ice/acetone mixture as required. The aluminium block containing the centrifuge tubes is removed from the dry-ice/acetone mixture after 64-88 hours and placed in the body of the deep-freeze where its temperature is allowed to rise gradually to -35°C and maintained for 2 hours. The following steps will then depend upon the resin and temperature chosen for final embedding and polymerisation as detailed in the protocols below.

4.4   Protocols for Cryosubstitution (Epoxy Resins)

4.4.1 Protocol 1: Epoxy Resins

4.4.1.1 Substitution Medium

The substitution medium (acetone or methanol) would normally contain a fixative, 0.1-2.5% osmium tetroxide, with if desired glutaraldehyde (1-3%) and/or uranyl acetate (0.1-0.5%) added. Osmium can be omitted in favour of just the aldehyde with or without uranyl acetate, or even no fixative used at all (Sect. 4.3.1). In the last case, the level of preservation that can then be achieved will be largely dependant upon the particular properties or chemical composition of the tissue being examined (Cryoimmobilisation and Resin Embedding in Sect. 1.3.2.2).

4.4.1.2 Substitution

The samples, following preparative procedures and freezing, are transferred to the substitution medium maintained at -80°C and left for 3-4 days. The temperature is raised to -35°C over a period of not less than 1 hour. After 2 hours at -35°C, the samples are placed at 0°C or 4°C for 2 hours, and finally at room temperature for 2 hours. If osmium is being used, the samples should be kept in the dark, where, even though the substitution medium darkens, this effect is acelerated by light.

4.4.1.3 Infiltration

At room temperature:

1.	Pure acetone (methanol)	30 min x 3
2.	1:2 neat resin*:acetone (methanol)	60 min
3.	1:1 neat resin:acetone (methanol)	60 min
4.	2:1 neat resin:acetone (methanol)	overnight (vial tops off)
5.	Neat resin*	120 min x 4
6.	Place tissue in resin-filled capsules	

*Neat resin refers to the final mixture of epoxy resin to which has been added the relevant accelerator (Glauert, 1975, 1991; Sect. 2.1.2).

4.4.1.4 Polymerisation

By heat at 60°C (Sect. 5.2.1).

4.5   Protocols for Cryosubstitution (Acrylic Resins)

4.5.1 Protocol 2: LR White

LR white can be used for infiltration and embedding from temperatures of -20°C to ambient.

4.5.1.1 Substitution Medium

As for epoxy resins (Protocol 1 above, Sect. 4.4.1.1; also see Sect. 4.3.1 for further discussion of these points, as osmium and other fixatives can be omitted from the substitution medium).

4.5.1.2 Substitution

The samples, following preparative procedures and freezing, are transfered to the substitution medium maintained at -80°C and left for 3-4 days. The temperature is raised to -35°C over a period of not less than 1 hour. After 2 hours at -35°C, the temperature is raised to -20°C for 2 hours. It is then raised to 0°C for 2 hours, and finally to room temperature for 2 hours. Processing in to resin can proceed directly from any of the last three temperature points using the same infiltration regimen.

4.5.1.3 Infiltration

At -20°C, 0°C, or room temperature:

1.	Pure acetone (methanol)	60 min
2.	1:1 neat resin*:acetone (methanol)	60 min
3.	2:1 neat resin:acetone (methanol)	60 min
4.	Neat resin*	60 min
5.	Neat resin	overnight
6.	Neat resin	60 min
7.	Place tissue in resin-filled capsules	

NOTE: Acetone is a free-radical scavenger; it must be completely exchanged with resin (Sect. 5.1)
*When using LR White, the hard-grade (or original) resin is recommended for cytochemistry and immunocytochemistry so as to preserve good sectioning qualities while minimising the extent of resin cross-linking during polymerisation (Chap. 5).

4.5.1.4 Polymerisation

At the particular temperature chosen either by heat, chemical or UV-light (Chap. 5).

4.5.2 Protocol 3: Lowicryls K4M and HM20

These low temperature resins can be used over a temperature range of -50°C to room temperature. For protocol steps in the range -20°C to ambient see the steps described for LR White (Protocol 2 above, Sect. 4.5.1). For -35°C or -50°C see below.

4.5.2.1 Substitution Medium

Acetone or methanol. No fixatives are added (see discussion under Sects. 1.3.2.2/4.3.1).

4.5.2.2 Substitution

The samples, following preparative procedures and freezing, are transferred to the substitution medium maintained at -80°C and left for 3-4 days. The temperature is raised to -35°C or -50°C over a period of not less than 1 hour.

4.5.2.3 Infiltration

At -35°C with either Lowicryls K4M or HM20. At -50°C with Lowicryl HM20 only.

1.	Pure acetone (methanol)	60 min
2.	1:1 neat resin*:acetone (methanol)	60 min
3.	2:1 neat resin:acetone (methanol)	60 min
4.	Neat resin*	60 min
5.	Neat resin	overnight
6.	Neat resin	120 min
7.	Place tissue in resin-filled capsules	

NOTE: Acetone is a free-radical scavenger it must be completely exchanged with resin (Sect. 5.1)
*Lowicryls K4M or HM20 complete with crosslinker and monomer (see Varying the Hardness of the Resins in Sect. 2.2.3.3).

4.5.2.4 Polymerisation

With UV-light at -35°C or -50°C (Sect. 5.4.2.3). Chemical polymerisation at -35°C is also possible (Sect. 5.3.2.4).

4.5.3 Protocol 4: Lowicryls K11M and HM23

Cryosubstitution and resin embedding are only possible down to -50°C with Lowicryl HM20. Distinct advantages may be gained by embedding tissue at even lower temperatures, although these have as yet still to be demonstrated (see Cryoimmobilisation and Resin Embedding under Sect. 1.3.2.2). Two new resins which are variations of Lowicryls K4M and HM20 (Acetarin et al, 1986) enable temperatures below -50°C to be reached. Lowicryl K11M can be UV-light polymerised at temperatures down to -60°C and Lowicryl HM23 down to -80°C. Both resins remain low in viscosity at these temperatures (Acetarin and Carlemalm, 1985; *N.B.* these resins can be used, if desired, at other higher temperatures as well).

Methods for freezing tissue remain the same (Sect. 4.2) and the substitution medium is either pure acetone or pure methanol. No fixative is added and the medium chosen is used as described in Protocol 3 (Sect. 4.5.2) for Lowicryls K4M and HM20 for 72-96 hours at -80°C. Little has been published on the use of Lowicryls K11M and HM23 for very low temperature embedding (Carlemalm et al, 1985b; Acetarin et al, 1986; Dürrenberger et al, 1988; Edelmann, 1989a, 1989b, 1991; Quintana, 1993; Villiger and Bremer, 1990), therefore, only tentative suggestions are offered (Hobot, 1989; Villiger, 1991; J.A.Hobot, unpublished observations). Protocol 4 should be viewed as a starting point to forming a consistent protocol for these very low temperature resins. Much longer times may be needed for infiltration and polymerisation, although excessive times could lead to loss of structure and tissue reactivity. Data will have to be accumulated by empirical means. It can be very difficult to maintain the low temperatures required for these long periods. Employing automatic equipment simplifies this problem (Sect. 4.3.3). Specialised commercial freezers with temperature ranges of -20°C to -90°C can be used for substitution and polymerisation at -60°C to -80°C (Merck; Fisons Scientific Equipment). Solutions are precooled to the relavent temperature before use. When using commercial freezers or cold mixtures, great care must be taken when opening the vials, as there is a grave and persistent danger of condensation with water/ice forming in the substitution medium or infiltrating resin.

4.5.3.1 Substitution Medium

Acetone or methanol. No fixatives are added (see discussion under Sects. 1.3.2.2/4.3.1).

4.5.3.2 Substitution

The samples, following preparative procedures and freezing, are transferred to the substitution medium maintained at -80°C and left for 3-4 days. The temperature is raised from -80°C to -60°C for Lowicryl K11M and kept at -80°C for Lowicryl HM23.

4.5.3.3 Infiltration

At -80°C for Lowicryl HM23, and at -60°C for Lowicryl K11M.

1.	Pure acetone (methanol)	2 h
2.	1:1 neat resin*:acetone (methanol)	8 h
3.	2:1 neat resin:acetone (methanol)	8 h
4.	Neat resin*	overnight

5. Neat resin   8 h
6. Neat resin   overnight (optional for some
                                            tissues)
7. Place tissue in resin-filled capsules

NOTE: Acetone is a free-radical scavenger it must be completely exchanged with resin (Sect. 5.1)
*Lowicryls K11M and HM23 complete with crosslinker and monomer (see Varying the Hardness of the Resins in Sect. 2.2.3.3).

4.5.3.4 Polymerisation

By UV-light at -60°C or -80°C (Sect. 5.4.2.4).

4.6 Freeze-Drying

Freeze-drying is dependant on the same methods of tissue preparation and rapid freezing as described for cryosubstitution (Sects. 4.1/4.2) but it has the advantage of avoiding organic solvents. The preservation of freeze-dried tissue, especially lipid/membrane barriers, may be different from that produced by cryosubstitution which could especially affect the infiltration of viscous Epoxy resins including Spurr's, but may be different with the true low viscosity resins like, for example, Lowicryl HM20 (viscosity similar to water at -35°C to -50°C). Lowicryl K4M, with a higher viscositiy than Lowicryl HM20, was found in one study to infiltrate tissues poorly at -35°C and so was used at -20°C (Chiovetti et al, 1986, 1897). It should though be used at -35°C, but if there are problems with infiltration then switching to Lowicryl HM20 could probably be the answer. Lowicryls K11M and HM23 at temperatures of -60°C to -80°C have infiltrated tissue well (Edelmann, 1986; Wroblewski et al, 1990), but with infiltration times of up to 24 hours. Protocols are, therefore, difficult to establish. Prolonged infiltration times may well be necessary. Whether the tissue is in reality "dry", and contains no water, may not be so important with acrylic resins because of their polar properties (Sect. 2.2.3.2). However, any water remaining in the tissue above -60°/-50°C can recrystalise and damage ultrastructure before resin infiltration occurs.

It is important to note that the Lowicryls, especially Lowicryl HM20, are volatile even at low temperatures, so that infiltration should *not* be conducted under vacuum or some resin components will evaporate. Any vacuum should be broken with dry nitrogen, and the neat resin immediately introduced to the tissue. Once infiltration is completed, polymerisation is by either UV-light or chemical catalytic methods at the desired temperature (Chap. 5).

Protocols which involve bringing the freeze-dried tissue samples up to room temperature followed by infiltration and embedding in resin, usually an Epoxy, initially involve the use of vapour fixation or stabilisation. The samples are exposed to fixative vapours (of osmium tetroxide or paraformaldehyde) under vacuum, after which the Epoxy resin is introduced prior to breaking the vacuum and then continuing the infiltration in sample vials. Exposure of tissue to vapour fixation from osmium tetroxde crystals for 15 min has provided reasonable fixation of pancreatic islets (Dudek et al, 1982, 1984) or smooth muscle (Chiovetti et al, 1987), but longer times (3-6 hours) have been employed for a variety of tissues, and immunocytochemistry performed (Linner et al, 1986; Livesey et al, 1989). However, pretreatment of the resin sections may be required to remove the osmium (Sect. 8.1.5), which will often reduce the immunoreactivity that can be retained within tissue (Sect. 1.2). Alternatively, paraformaldehyde vapours can be employed (Dudek et al, 1984). The drawbacks of using vapour fixation and resin embedding at room temperatures have already been discussed earlier (in Freeze-Drying under Sect. 1.3.2.2). A positive step in avoiding such drawbacks would be to start resin infiltration at low temperatures (i.e. below -30°C) using one of the Lowicryls.

4.7   Protocols for Freeze-Drying (Epoxy Resins)

4.7.1 Protocol 1

After the freeze-drying cycle has been completed and stopped at room temperature the tissue samples are processed by vapour fixation and resin infiltration.

4.7.1.1 Vapour Fixation

0.25 g of osmium tetroxide are introduced, under vacuum, in the freeze-drying apparatus for 15 min to 3 hours. Alternatively, paraformaldehyde is added to the apparatus chamber (each particular apparatus will have its own special pre-pumped chambers, holders/wells for samples and fixatives) for 1-3 hours. Shorter times may be employed, but deduced empirically for any particular tissue type.

4.7.1.2 Resin Infiltration

The de-gassed resin is introduced under vacuum, and after 1 hour the vacuum is broken and infiltration proceeds at room temperature in specimen vials on a rotary device.

1.   Neat resin*   60 min

2.	Neat resin	overnight
3.	Neat resin	8 h
4.	Neat resin	overnight (optional, depending upon if tissue is fully infiltrated)

*Neat resin refers to the final mixture of resin to which has been added the relevant accelerator (Glauert, 1975, 1991; Sect. 2.1.2).

4.7.1.3 Polymerisation

By heat at 60°C (Sect. 5.2.1).

4.8 Protocols for Freeze-Drying (Acrylic Resins)

The freeze-drying cycle can be completed at the lowest temperature at which the particular acrylic resin is still usable: -80°C, Lowicryl HM23; -60°C, Lowicryl K11M; -50°C, Lowicryl HM20; -35°C, Lowicryl K4M; -20°C, LR White; 0°C up to room temperature for all. Below -30°C it is not necessary to use vapour fixation.

4.8.1 Protocol 2: 0°C to Room Temperature

4.8.1.1 Vapour Fixation

As set out for the Epoxy resins (Protocol 1 above, Sect. 4.7.1.1). In some cases it may be possible to avoid this step (see discussion under Sect. 1.3.2.2).

4.8.1.2 Resin Infiltration

The infiltration times are as set out for the Epoxy resins (Protocol 1 above, Sect. 4.7.1.2) and can be applied to LR White or any of the Lowicryl resins. However, *N.B.:* Acrylic resins are not de-gassed. They are generally most volatile, and could lose some components during de-gassing. They are therefore not introduced in to the freeze-drying apparatus under vacuum (Sect. 4.6). The vacuum is broken by the introduction of dry nitrogen gas and the (pre-cooled) resin added.

When using LR White, the hard-grade (or original) resin is recommended for cytochemistry and immunocytochemistry so as to preserve good sectioning qualities while minimising the extent of resin cross-linking during polymerisation (Chap. 5).

Lowicryls K4M or HM20 are complete with crosslinker and monomer (see Varying the Hardness of the Resins in Sect. 2.2.3.3).

4.8.1.3 Polymerisation

By heat, chemical catalytic methods or UV-light (the last only if osmium was not used) as set out in Chap. 5.

4.8.2 Protocol 3: -20°C to -80°C

4.8.2.1 Vapour Fixation

None is required (see discussion under Sect. 1.3.2.2).

4.8.2.2 Resin Infiltration

The vacuum is broken with dry nitrogen gas and the pre-cooled resin immediately introduced (Sect. 4.8.1.2).
    After 1 hour the tissue samples are processed at the particular temperature chosen.

1.	Neat resin*	60 min
2.	Neat resin	overnight
3.	Neat resin	2 h
4.	Neat resin	8 h (optional, depending upon if tissue is fully infiltrated)
5.	Neat resin	overnight (optional, depending upon if tissue is fully infiltrated)

*Lowicryls K4M or HM20 complete with crosslinker and monomer (see Varying the Hardness of the Resins in Sect. 2.2.3.3).
*Lowicryls K11M and HM23 complete with crosslinker and monomer (see Varying the Hardness of the Resins in Sect. 2.2.3.3).

4.8.2.3 Polymerisation

By chemical catalytic methods or UV-light for LR White and Lowicryls K4M or HM20; by UV-light for Lowicryls K11M and HM23 (see Chap. 5 for full details).

4.9   Cryoultramicrotomy

The methodologies presented here are intended as a brief guide, for comparison with the main theme of this book which is resin embedding. Cryoultramicrotomy may provide advantages in areas where organic solvent or organic resins adversely affect the retention of antigenicity within tissue although little has been reported of such examples. Tissue fixation is still an essential prerequisite which means that cryoprotection can also be used. The protocols set out below are to be found in Tokuyasu (1973, 1984, 1986) or Hobot and Newman (1991). A useful introduction to cryoultramicrotomy may be found in Reid and Beesley (1991).

4.9.1 Fixation

Tissue can be fixed by vascular perfusion or immersion for 15-60 min with glutaraldehyde (0.5-1%) or glutaraldehyde/formaldehyde mixtures. To be borne in mind are (a) the limitations in retaining antigenic sensitivity that such levels of fixation offer (Sect. 1.3.2), and (b) the relative merits of perfusion versus immersion to the integrity of the cyosections once "thawed" (discussed in Cryoultramicrotomy for Immunocytochemistry in Sect. 1.3.2.2). Lowering the fixation concentrations to reach the sensitivity and quality of structural preservation that can be retained by resin embedding procedures (e.g. partial dehydration, PLT; or no fixation by cryosubstitution), may not be possible with cryoultramicrotomy and is an area that needs close examination (Hobot and Newman, 1991; Cryoultramicrotomy for Immunocytochemistry under Sect. 1.3.2.2).

Following fixation small pieces (1 mm^3) of tissue are infiltrated with sucrose solutions (0.6-2.3M) made up in the appropriate buffer (times will vary depending upon tissue type and block size). The presence of sucrose has two advantages: firstly it acts as a cryoprotectant; and secondly, it improves the sectioning qualities of the frozen tissue block.

4.9.2 Apparatus Requirements

Freezing of tissue is carried out by plunging or dropping the specimens in to a Dewar containing liquid nitrogen or freon. The specialised equipment listed in this chapter (Sect. 4.2) can also be used if desired, but cryoprotection lessens the need for such critical apparatus.

The tissue is sectioned whilst still frozen, and therefore a cryo-attachment is necessary for the ultramicrotome. Two main suppliers of such equipment are Leica (Reichert FC S Low Temperature Sectioning System) or RMC (CR-21 Cryosectioning Unit). Both can be fitted on to the manufacture's ultramicrotome models, although the RMC attachment can be easily adapted to several makes of ultramicrotome. It also requires a smaller Dewar of liquid nitrogen to be positioned in the laboratory (10 L vs. 35 L for the Leica model, although the Leica Dewar and liquid nitrogen pump can be employed with their rapid freezing apparatus as well). For both, the cryo-attachments slip on and off the ultramicrotome very simply and quickly.

4.9.3 Sectioning and Storage of Grids

Sectioning is carried out on the cryoultramicrotome, and has similar problems to those associated with resin sectioning. (True cryosectioning i.e. of unfixed, non-cryoprotected tissue to produce frozen-hydrated sections, has other problems but these are not for discussion here). The sections are picked up on a drop of the above sucrose solution held in a wire loop, touched on to the surface of a nickel grid (300 mesh, hexagonal, high transmission) having a carbon-coated collodion film, and transferred section side down to a wet plate of 0.3% agarose/1% gelatin (made up in 0.01M PBS). This ensures that the sucrose drop diffuses slowly away from the grid without the sections falling off. The grids are then ready for immunolabelling (Sect. 9.5). Frozen blocks still wanted for further sectioning may be stored under liquid nitrogen (Cryoultramicrotomy for Immunocytochemistry in Sect. 1.3.2.2).

5   Methods for Resin Polymerisation

5.1 Introduction

Heat, chemical catalysts and UV-light are the most popular forms of polymerisation for current EM resin systems. LR Gold is polymerised by blue-light. Epoxides can only be cured by heat to produce sectionable blocks. It is possible but impracticable to cure Epoxy resins with UV-light or chemical catalysts (Glauert, 1975). However, considerably longer times are needed for the tissue to be present in an organic monomer (Sect. 1.3.1). These methods are not used routinely and references to them in the literature are scarce.

Acrylic resins can be polymerised by all the listed procedures. In addition, it should be noted that the polymerisation of acrylic resins is adversely affected by the presence of either oxygen or acetone. Both act as radical scavengers, lowering the amount of free radicals produced during the polymerisation process, which will reduce the cross-linking of resin polymers (sect. 1.3.4). Care should, therefore, be taken in handling or mixing acrylic resin components, and if acetone is used as a dehydrating solvent, then a complete exchange with resin should occur prior to polymerisation. When flat embedding acrylic resins (see also Glauert and Young, 1989), care must be taken in devising a system that excludes air containing oxygen. The introduction of a dry nitrogen atmosphere is helpful.

Gelatin capsules are recommended for holding the tissue and acrylic resin during the embedding stage, because gelatin completely excludes air, whereas some BEEM-type capsules do not. The polymerisation of acrylic resins is an exothermic reaction and, therefore, to minimise the evolution of heat it is essential that the volume of resin to be used for polymerisation is kept to a minimum (i.e. do not employ larger gelatin capsules than size '0'/vol. 0.68 ml). Gelatin melts at about 70°C and softens at lower temperatures, and so acts as a useful temperature indicator. If there is distortion in the capsules shape this could indicate that high temperatures have been reached. To avoid any adverse effects of heat production during acrylic resin polymerisation, especially by heat or chemical catalysts, it is *strongly recommended* that the gelatin capsules are placed in a metal heat sink. An aluminium block drilled to accommodate the capsules will act as an efficient heat distributor, evening out any temperature rise as a result of the exotherm (Acetarin and Carlemalm, 1982).

An important compound used in the polymerisation of acrylic resins is the catalyst dibenzoyl peroxide. It is already present in all of the three LR White grades (soft, medium, hard), with the 'hard grade' product being used for all the described procedures in the book. The polymerisation of the recently introduced 'uncatalysed' LR White to which a 'benzoyl peroxide powder catalyst', as supplied by the manufacturers, has to be added is dealt with in the last section of this chapter. For heat or chemical catalytic polymerisation, the Lowicryl formulations are made up of crosslinker and monomer, to which dibenzoyl peroxide has to be added (see Varying the Hardness of the Resins in Sect. 2.2.3.3). Extreme caution should be exercised when using dibenzoyl peroxide as when dry it is highly explosive. It is therefore supplied as a hydrated product, moistened with 25% water (Agar Scientific). It can be dried following competent chemical procedures, or it can be used in its hydrated state but the actual amount of dibenzoyl peroxide present being calculated as composing 75% of the total weight of the product supplied.

The suggestions for heat and chemical catalytic polymerisation of Lowicryls K4M and HM20 form only a basis for experimentation. The proportions of all the resin components can be changed empirically, producing polymers with quite different properties (Sect. 5.3.2.5). These forms of polymerisation have not been popular so that few data are available on their application (Hobot, 1990). They were introduced to cope with pigmented specimens which could possibly adsorb UV photons and lead to incomplete UV-light polymerisation (Acetarin and Carlemalm, 1982). In addition, other factors such as the tissue type and whether or not it has been osmicated or "block-stained" with uranyl acetate will alter the characteristics of the polymerised blocks. For these reasons ultra-violet (UV) light polymerisation has been preferred.

An important point concerning the chemical polymerisation of acrylic resins is that once the appropriate amounts of catalyst and accelerator/activator have been added, then polymerisation, i.e. polymer cross-linking, will proceed to completion. At this stage the resin block is very likely to be of a hardness which would make sectioning most difficult. To overcome this problem, the procedures for chemical polymerisation of the acrylics stress the importance of storing under-polymerised resin blocks at low temperatures (-20°C to -40°C) following polymerisation for a given time (Sect. 6.2.2). Leaving the resin blocks on the bench at ambient temperatures for even 1-3 hours can result in further undesirable hardening of the blocks. The importance of being able to use chemical catalytic polymerisation to control the final cross-linked density of acrylic resins is discussed elsewhere (Sects. 1.3.4/2.2.1.3 and Chap. 10).

5.2   Heat Polymerisation Methods

5.2.1 Epoxy Resins

The resin infiltrated tissue blocks are left in resin filled '0'-guage (or smaller) gelatin capsules containing paper labels for 16-36 hours at 60°C depending on the resin (Causton, 1980). As no exothermic reaction is involved in the curing process, and nor is it affected adversely by oxygen, larger embedding moulds or BEEM-type capsules can be used.

5.2.2 Acrylic Resins

5.2.2.1 LR White

'0'-guage (or smaller) gelatin capsules containing paper labels are fully-filled with recently purchased LR White resin. It should be less than three months old, and stored at 4°C. Older resin can be used for resin exchanges during infiltration. The tissue pieces, with minimum carry-over of resin, are quickly transferred in to capsules, which are then tightly capped and preferably placed in an aluminium block. Polymerisation is conducted at exactly 50°C in an incubator for 24 hours. Slow heat curing is believed to create a linearity of molecular structure which is conducive to penetration of sections by aqueous reagents (B.Causton, personal communication). Although LR White is inhibited from polymerising by the presence of oxygen, the tiny amount of air trapped in the lid of a gelatin capsule only leads to a slight stickiness on the end of the block distal to the tissue. The resin is by no means fully polymerised at this stage. Further polymerisation can be achieved by simply leaving the block(s) at 50°C for a longer time. However, blocks are often sufficiently cured by the initial polymerisation, so it is advisable to thin-section one and test for electron beam stability before cautiously continuing polymerisation, otherwise the blocks may become too brittle. Blocks which have been opened for sectioning are difficult to polymerise further so some should always be kept unopened in reserve if possible. Blocks can be stored as detailed in Sect. 6.2.2.1

5.2.2.2 Lowicryls

The Lowicryl K4M and HM20 kits are designed for low temperature use. Therefore, in order to polymerise the resins with heat, dibenzoyl peroxide (DBP; Agar Scientific) must be added in place of benzoin methyl ether. 0.6% by weight (total; actual weight of DBP is 0.45%) should be added to Lowicryl K4M and Lowicryl HM20 which will provide sufficient free-radicals to polymerise the resins in 24-48 hours at 50°C (recommended temperature for heat polymerisation). Lowicryl HM20 is less reactive than Lowicryl K4M (Acetarin and Carlemalm, 1982), and therefore the amount of DBP can be increased to 0.8% (total; actual weight of DBP is 0.6%). For 60°C, the same amounts of DBP are used. Gelatin capsules (size '0' or smaller) should only be three-quarters filled to allow for initial swelling of the resin, and are placed in an aluminium block for polymerisation. The mixture of resin and DBP is used for the infiltration steps. Blocks can be stored as detailed in Sect. 6.2.2.2.

5.3   Chemical Catalytic Polymerisation Methods

5.3.1 LR White

5.3.1.1 Room Temperature

The tissue blocks are dropped in to '0'-guage (or smaller) gelatin capsules (containing paper labels) that have been fully-filled with recently purchased LR White (< 3 months old and kept in a refrigerator at 4°C) to which has been added the manufacturer's accelerator (also recently purchased and kept as for the resin; accelerator is equivalent to activator; Sect. 1.3.4) in the ratio of 1.5 µl of accelerator to 1 ml of resin. The resin and the accelerator are used directly from the 4°C refrigerator and should be stirred together (30 secs) with a clean glass rod or an orange stick (i.e. thin, wooden applicator sticks) avoiding the formation of bubbles. Air introduced in to the mixture will inhibit polymerisation. At 22°C the mixture will take 20-30 minutes to completely harden but may start to gel after as little as 7 minutes. As a result of this short working period it is advisable to deal with only 5 or 10 blocks at a time using 5 or 10 ml aliquots of LR White. Tissue blocks are placed on to the surface of the resin at the top of the gelatin capsules and allowed to sink to the bottom. It is also important to minimise carry-over of resin from the infiltration steps. The capsules are quickly capped.

The polymerisation reaction is exothermic and results in the temperature rising from 22°C to 50-60°C in the gelatin capsules. This rise is transient and after a few seconds rapidly falls again (Yoshimura et al, 1986). It is not necessary to cool the blocks. Not all the monomer polymerises, as the volume of accelerator added is a carefully monitored amount to produce blocks of sufficient hardness for sectioning, but which are not fully cross-linked (Newman and Hobot, 1987; Yoshimura et al, 1986; Sect. 1.3.4). About 10% of the resin in the end of the capsule distal to the tissue should remain liquid. If it does not then the LR White was too old or was incorrectly stored. If the LR White remains completely unpolymerised oxygen may have been introduced slowly in to the stock solution, or the accelerator has degenerated (was too old or badly stored). In these cases, the tissue should be removed and re-embedded in fresh resin/catalyst mixture. The large volume of unused resin, remaining in the vials after filling the capsules, will become very hot, often emitting smoke-like fumes, as it polymerises. This high temperature is not reached by the much smaller volume of LR White in the gelatin capsules (Yoshimura et al, 1986). An aluminium block can be used as a heat sink.

As with heat polymerisation, the blocks are less than completely cross-linked after the initial polymerisation but the polymerisation of unopened blocks can be continued, to increase the electron beam stability of sections cut from them, by placing the blocks at 50°C. One hour of further polymerisation is often enough to produce thin sections that are viewable in the electron microscope unsupported on 300 mesh high-transmission grids. To prevent further cross-linking of resin taking place, the blocks can be stored, prior to the 50°C step, at temperatures of -20°C or lower (Sect. 6.2.2.1).

5.3.1.2 Cold Temperature (0°C)

5 or 10 ml aliquots of fresh, recently purchased resin are precooled to 0°C. '0'-guage (or smaller) gelatin capsules, containing paper labels, are made ready in a suitable holder, for example an aluminium block precooled to 0°C. As soon as an aliquot of resin is removed from the refrigerator it is mixed with the manufacturer's accelerator (also recently purchased and correctly stored) in the ratio of 1.5 µl accelerator/1 ml of resin, i.e. 7.5 µl accelerator in a 5 ml aliquot of resin or 15 µl in a 10 ml aliquot of resin. The acelerator should be rapidly stirred (30 secs) in to the resin with a clean glass rod or orange stick but air-bubbles must be avoided. It is helpful to keep the resin/accelerator mixture on ice (0°C). It is dispensed rapidly in to the awaiting gelatin capsules (preferably held in the aluminium heat-sink, see Sect. 5.1). Tissue blocks are placed in to the surface of the resin at the top of the gelatin capsules and allowed to sink to the bottom. It is also important to minimise carry-over of resin from the infiltration steps. The capsules are quickly capped and returned to the 0°C refrigerator. It is preferable to work with small amounts of resin, embedding only 5 or 10 blocks in batches, so that the resin mixture does not have time to warm appreciably, particularly in the absence of ice or a cooled aluminium block. The more rapidly the blocks are returned to the refrigerator, the more controlled will be the polymerisation. The resin/catalyst mixture, because of the longer time it takes to gel at 0°C (2-4 hours), has more time to penetrate to the centre of the tissue blocks resulting in a more even polymerisation.

It is probably most convenient to leave blocks at 0°C overnight because polymerisation takes 3-4 hours. 25-30% of the resin should remain liquid in the end of the capsules distal to the tissue. At this stage, the resin is minimally crosslinked. Semithin sections can easily be cut and the tissue is very reactive, staining heavily with routine stains and having considerably more sensitivity to immunoreagents and lectins than tissue embedded in epoxide or in LR White by heat or room temperature chemical catalytic polymerisation. However, thin sections are susceptible to damage by the electron beam and will need the support of a carbon or plastic film. Thin sections that are stable in the electron beam and can, therefore, be used unsupported, can only be cut from blocks that have been further polymerised for 1-2 hours at 50°C. The liquid resin in the ends of these blocks usually polymerises during this process. There is no appreciable loss in the tissue's immunoreactivity, and low concentration, secondary sites (if partial dehydration has been used; Sect. 3.3.1.2/3.3.3) still immunolabel well (Hobot and Newman, 1991; Newman, 1989; Newman and Hobot, 1989). As above (Sects. 5.2.2.1/5.3.1.1), complete polymerisation of the LR White should be avoided because the blocks become hard and brittle and the tissue unresponsive to immunolabels and counterstains. It is expedient to embed sufficient blocks of tissue to allow some to be used for semithin sectioning immediately following initial polymerisation (taking advantage of their high sensitivity) and for others to be further polymerised for thin sectioning. Blocks which have already been sectioned have a limited repolymerisation potential. To preserve the immunoreactivity of the blocks, they can be stored at temperatures of -20°C or lower before carrying out any further polymerisation at 50°C (Sect. 6.2.2.1).

5.3.1.3 Cold Temperatures (-20°C)

The procedure and precautions taken are the same as for 0°C (Sect. 5.3.1.2) except that the amount of accelerator added can be increased up to 3 µl per 1 ml of resin. Polymerisation is at -20°C, and blocks can be stored at -40°C (Sect. 6.2.2.1).

5.3.2 Lowicryls

The formulations given for chemical catalytic polymerisation of the Lowicryls for room temperature to -20°C are calculated for providing a similar schedule in handling and polymerisation to that obtaining with LR White. Both resins should be recently purchased (3-6 months), this being especially more important for the less reactive Lowicryl HM20. *Note the weights of dibenzoyl peroxide being added* (Sect. 5.1). For experimental purposes, for example to vary the cross-linked density of the resin (Sect. 1.3.4; Chap. 10), formulations derived from the work of Acetarin and Carlemalm (1982) are given in a separate section below (Sect. 5.3.2.5).

5.3.2.1 Room Temperature

In addition to dibenzoyl peroxide (DBP), it is necessary to add an activator/accelerator, N,N-dimethylparatoluidine (DMpT; Fluka). DMpT is an amine compound, and being toxic, should be handled with great care. For room temperature polymerisation of Lowicryl K4M, 0.6% DBP (total weight; actual weight of DBP is 0.45%) and 0.15% DMpT, by weight, are added. An accurate way in which to prepare the complete resin mix is to first combine the cross-linker and resin monomer (Varying the Hardness of the Resin in Sect. 2.2.3.3). To this is added the catalyst, DBP, and the resin mixture used for the infiltration steps. For actual polymerisation, the recommended amount of DBP (60 mg total weight) is added to 10ml of pure resin, and just prior to placing the tissue blocks in the gelatin capsules, the appropriate amount (15 µl) of DMpT is added. As before, whenever mixing acrylic resins avoid introducing air. Mixing of DBP is best achieved by bubbling dry nitrogen through the resin solution, and for DMpT by gently stirring for 30 seconds using a clean glass rod or orange stick. The resin mixture is rapidly dispensed in to '0'-gauge (or smaller) gelatin capsules in to which are also added the tissue blocks and labels. The resin will begin to gel after 7-8 minutes, so there is a need to work quickly. Tissue blocks are placed in to the surface of the resin at the top of the gelatin capsules and allowed to sink to the bottom. It is also important to minimise carry-over of resin from the infiltration steps. The capsules are quickly capped. The gelatin capsules are placed in an aluminium block for polymerisation (Sect. 5.1). Blocks are hardened after 30 min in the case of Lowicryl K4M with the top 10-25% liquid.

Lowicryl HM20 is less reactive than Lowicryl K4M (Acetarin and Carlemalm, 1982) and consequently more of the catalyst/activator are required. To the complete reisn mixture (Varying the Hardness of the Resins in Sect. 2.2.3.3) 0.8% of DBP (total weight; actual weight of DBP is 0.6%) and 0.6% of DMpT are added. Polymerisation time is 30 min.

Handling and storage of the blocks is the same as for LR White (see above, Sect. 5.3.1.1 and Sect. 6.2.2.2).

5.3.2.2 Cold Temperature (0 °C)

The same formulations as for room temperature polymerisation (Sect. 5.3.2.1) are used. However, after addition of DBP, the resin to be used for polymerisation is precooled to 0°C and then the appropriate amount of DMpT added. (If for the less reactive Lowicryl HM20 polymerisation times are prolonged, then it may be necessary to increase the amounts of DMpT added to 0.8%). The resin is gently stirred (30 secs) with an orange stick to disperse the DMpT and dispensed in to small gelatin capsules ('0'-gauge or smaller) sitting in a precooled aluminium heat/cold sink. Tissue blocks are placed in to the surface of the resin at the top of the gelatin capsules and allowed to sink to the bottom. It is also important to minimise carry-over of resin from the infiltration steps. The capsules are quickly capped. Polymerisation proceeds for 24 hours at 0°C, although blocks may be already hardened after 3-4 hours. The top 10-25% may be liquid and the blocks are handled in the same way as set out for LR White (Sect. 5.3.1.2). They can be stored at -20°C or lower to prevent further polymerisation (Sect. 6.2.2.2). The resin blocks can be placed for 1-2 hours in an incubator at 50°C to improve their sectioning qualities and the electron beam stability of thin sections, with no adverse affect on ultrastructure or a significant decrease in immunoreactivity within the tissue (Newman, 1989; Newman and Hobot, 1987). When working on the bench it is recommended that the above procedures are carried out on ice. The mixture of resin plus DBP is used for infiltration.

5.3.2.3 Cold Temperatures (-20°C)

The procedure and precautions taken are the same as for 0°C (Sect. 5.3.2.2), except that the amount of accelerator added can be increased up to 3 µl per 1 ml of resin in the case of Lowicryl K4M. With the less reactive Lowicryl HM20, the amounts of DBP and DMpT can be increased empirically (see Sect. 5.3.2.5) to perhaps 1% DBP (total weight; actual weight of DBP is 0.75%) and 0.8% DMpT. Polymerisation is at -20°C for 24 hours, and blocks can be stored at -40°C (Sect. 6.2.2.2).

5.3.2.4 Low Temperatures (-35°C)

For polymerisation at -35°C, 0.53% DBP (total weight; actual weight of DBP is 0.4%) and 0.25% DMPT, by weight, should be added to Lowicryl K4M; and 1.33% DBP (total weight; actual weight of DBP is 1%) and 0.8% DMPT, by weight, should be added to Lowicryl HM20 (Acetarin and Carlemalm, 1982).

5.3.2.5 Experimental Parameters for Chemical Catalytic Polymerisation

In order to produce resin blocks which are still sectionable but have a reduced cross-linked density of the resin, the amounts of catalyst (DBP) and activator (DMpT) can be varied (Sect. 1.3.4; Chap.10). The formulations for varying the amounts of each are derived from the work of Acetarin and Carlemalm (1982), where the ratio of activator to catalyst is important in deciding the time required for gelation of the resin mixture to occur, the temperature at which polymerisation is to proceed, and the time for polymerisation to be completed. This ratio for activator to catalyst is approximately 0.6 to 0.625 (Acetarin and Carlemalm, 1982). Based upon this ratio and empirically derived results, some suggestions for the Lowicryls can be made. As it was important to examine the comparative merits of room temperature partial dehydration and PLT for retention of immunoreactivity within tissue (Hobot and Newman, 1991), studies on chemical polymerisation and the Lowicryls have mainly centred on utilising Lowicryl K4M.

  The formulation, for blocks which are very under-polymerised can be used at either room temperature (polymerisation time 90 min; approximately 50% of liquid resin at the top), or at 0°C (polymerisation time 24 hours, though blocks are hardened after 3-4 hours, with approximately 75% of liquid resin at the top). 0.117% DBP (total weight; actual weight of DBP is 0.088%) and 0.055% DMpT are added to Lowicryl K4M. For polymerisation at 0°C the resin plus DBP is precooled prior to the addition of DMpT (Hobot, 1990). The resin mixtures are handled in the same way as detailed above. Resin blocks of Lowicryl K4M polymerised in this way at 0°C following partial dehydration (Hobot, 1990) have shown the same novel structural organisation of the bacterial cell wall as seen after using LR White and cold catalytic polymerisation (Newman and Hobot, 1987) or PLT and the Lowicryls (Hobot et al, 1984). The advantages to be gained for improving tissue immunocytochemistry by reducing the cross-linked density of the resin are still to be elucidated experimentally, although the penetration of markers in to a less cross-linked resin meshwork may well increase the levels of detection above those currently obtainable (Chap. 10).

Under-polymerised blocks of Lowicryl HM20 can also be produced by for example using 0.6% DBP (total weight; actual weight of DBP is 0.45%) and 0.3% DMpT. Polymerisation time at room temperature is 60 min, with approximately 50% of the block hardened. It is important to store such blocks at low temperatures, as polymerisation will proceed to give completely hard blocks after 3 hours. The ratio of activator to catalyst can be lowered empirically to reduce the chances of polymerisation proceeding further than required, but only at the expense of prolonging polymerisation times.

5.4   Ultraviolet Light Polymerisation Methods

5.4.1 The Setting-up of Apparatus

The Lowicryl resins can be polymerised with UV-light at temperatures varying between 30°C and -35°C for Lowicryl K4M or as low as -50°C for Lowicryl HM20. A simple apparatus can be constructed to facilitate indirect UV illumination which is necessary initially to obtain even polymerisation throughout the tissue blocks. A small, sturdy cardboard box, that can be placed in a deep-freeze for low temperature embedding, is lined with aluminium foil shiny side out. A slot is cut in the lid to allow the UV source to be positioned. Two 15-W fluorescent tubes (Philips TLD 15-W/05) should be sited so that they are 30-40 cm above the resin-filled capsules containing the specimens. The wavelength of the light must always be 360 nm. Two 6-W UV lamps (Agar Scientific) can be used but then the distance between the lamps and the specimens must be reduced to 22-24 cm. This lower wattage apparatus generates much less heat and is useful for incorporating in to a -40°C deep-freeze where it does not affect the internal temperature of the cabinet (Hobot, 1989). In the majority of cases, unless the deep-freeze is very efficient, the use of two 15-W lamps could cause an appreciable rise of temperature within the freezer; in this situation the lamps will need to be mounted externally in relation to the freezer. Indirect illumination is achieved by having a right-angle deflector lined with shiny aluminium foil situated just below the UV source. '0'-gague (or smaller) gelatin capsules are small enough in volume to prevent an adverse rise in temperature as a result of the exotherm during polymerisation (the temperature rise is less than 2°C at -35°C; Weibull, 1986) and clear enough to allow penetration of the UV-light. Gelatin is also impervious to oxygen which seriously inhibits polymerisation. It is unwise to include labels which could occlude UV-light and the capsules should be held in such a way that the resin and tissue receive uninterrupted illumination, for example using thin wire loops as stands. UV polymerisation is easily performed at room temperature with the same lay-out.

Commercially available UV cabinets can be purchased, and these are capable of reaching temperaures of -35°C (UVF 35 Cold Cabinet, Agar Scientific) or -50°C (TTP 010 Low Temperature Polymerisation Apparatus, Balzers Union). Balzers Union also supplies an apparatus for use at room temperature (RTP 010). For very low temperatures of -60°C to -80°C, specialised commercial low temperature cabinets are available (Merck; Fisons Scientific Equipment) and can be adapted for use with the set-up outlined above for -35°C. The CS Auto from Leica can also reach these very low temperatures for polymerisation (Sects. 3.5.1/4.3.3).

The heat produced during the exothermic polymerisation reaction can, at room temperature, result in a brief temperature rise up to +12°C if the gelatin capsules are suspended in air (Weibull, 1986). There is a substantial reduction of the rise in temperature, to +4°C, if a heat sink is used (e.g. an alcohol bath; Glauert and Young, 1989; Weibull, 1986). At low temperatures, -35°C, and with the gelatin capsules susupended in air, the heat rise is small, less than +2°C (Glauert and Young, 1989; Weibull, 1986). It is very important (a) to keep the volume of resin small (< 0.7ml, '0' gauge), even for flat embeddings, (b) to start the UV-light polymerisation using *indirect* illumination of the specimens, and (c) to vary the distance of the light source to the specimens depending on the wattage of the UV-lamps (Sect. 1.3.4). Aluminium blocks should not be used as heat sinks for UV polymerisation, as this can lead to uneven and insufficient hardening of the resin (Glauert and Young, 1989). The spacing, if any, between gelatin capsule and the aluminium block's sides are much too small to allow adequate reflection of UV-light.

5.4.2 LR Resins and Lowicryls

Both classes of resin can be polymerised in exactly the same way by UV-light in the temperature range covering ambient temperatures to 20°C. The Lowicryls, however, can be polymerised at still lower temperatures, i.e. -20°C to -80°C. The polymerisation time, using initially indirect illumination, is 24 hours, and this is sufficient to produce sectionable blocks. (The blocks are already hard after 4 hours). Sectioning qualities are improved by a further 72 hours of illumination using a direct UV-light source. The initiator is dissolved in the resin by bubbling dry nitrogen through the mixture. Infiltration steps are carried out with the initiator already added to the resin of choice. The initiator must be added to LR Gold, *but is not necessary for LR White*.

5.4.2.1 Room Temperature

The initiator, benzoin methyl ether (reagent C), supplied with the Lowicryl kits is intended for low temperature polymerisation (-35°C to -50°C) and must be replaced with benzoin ethyl ether (Fluka) used in the same proportion i.e. 50 mg per 10 g resin. The resin mix is transferred to '0'-guage (or smaller) gelatin capsules and the small tissue blocks are placed on to the surface of the resin at the top of the capsules and allowed to sink to the bottom. Polymerisation should be by indirect UV-light for 24 hours, followed by direct UV-light for 48-72 hours. For polymerisation commencing at room temperature, the use of a heat sink (Sect. 5.4.1) can be recommended for lightly fixed specimens processed by partial dehydration (Sect. 3.3). The method is suitable for LR White, LR Gold and all the Lowicryls.

5.4.2.2 Cold Temperatures (0°C to -20°C)

For temperatures in the range 0°C to -10°C, use 50 mg of the initiator benzoin ethyl ether per 10 g of resin; for temperatures below -10°C use 50 mg of initiator C (benzoin methyl ether) per 10 g resin. When polymerising with UV-light the tissue blocks in resin-filled gelatin capsules are equilibrated at the chosen temperature (0°C to -20°C) for 10 min, then indirectly illuminated with UV-light at the chosen temperature for 24 hours, followed by direct illumination at room temperature for a further 72 hours. A heat sink may be used, but will probably not be necessary (Sect. 5.4.1.). The method is suitable for LR White, LR Gold and all the Lowicryls.

5.4.2.3 Low Temperatures (-35°C to -50°C)

For this polymerisation schedule the initiator supplied, benzoin methyl ether, is used in the proportions of 50 mg per 10 g resin. Illumination is with indirect UV-light at -35°C for 24 hours, and then with direct UV-light at room temperature for 72 hours. The method is suitable for Lowicryl K4M only down to -35°C and for Lowicryl HM20 down to -50°C. The use of a heat sink will probably not be necessary (Sect. 5.4.1.).

5.4.2.4 Very Low Temperatures (-50°C to -80°C)

For UV light polymerisation down to -60°C, 50 mg of initiator C (benzoin methyl ether) is added to 10g of Lowicryl K11M. Polymerisation is by indirect UV-light at -60°C for 3 days, followed by direct UV-light at room temperature for 2-3 days. The use of a heat sink at these very low temperatures is not essential.

The same proportions of initiator C and Lowicryl HM23 are used for temperatures down to -50°C. But for temperatures below -50°C and using Lowicryl

HM23, photoinitiator "J" (benzyl dimethyl acetal) should be used. Add 50-75 mg for temperatures of -50°C to -70°C, and 75-100 mg for -70°C to -80°C per 10 g of resin. Polymerisation is by indirect UV-light for 6 days, followed by direct UV-light at room temperature for 2-3 days.

At low temperatures polymerisation is slow, great care must be taken when mixing the resin components together so as not to introduce oxygen from the atmosphere which will inhibit polymerisation. Mixing is best achieved by bubbling dry nitrogen through the resin solutions. For both resins if polymerisation is not successful or incomplete (i.e. the resin has gelled but not hardened) after the initial illumination, it is best to proceed with UV polymerisation using (a) direct illumination at the appropriate low temperature, or (b) indirect UV-light at a higher temperature (e.g. -35°C) prior to room temperature illumination. Alternatively, it may be necesary to repeat the experiment but paying extra special attention to the preparation of the resin mixture, and perhaps contemplating starting with direct UV-light illumination if the indirect source is not successful.

5.5 'Uncatalysed' LR White

The presence of a catalyst in the original manufacturer's product can cause problems for orders placed in warm or hot climates. The resin can often arrive as a viscous liquid, already in the first stages of polymerisation. To overcome these handicaps, and to avoid shipping in dry ice, the alternative solution of omitting the catalyst from the resin mixture has been adopted. The catalyst can therefore, upon request, be supplied separately from the LR White which is now designated as 'Uncatalysed' LR White. The catalyst is supplied in 4.5 g amounts in small plastic vials as "LR White Catalyst Benzoyl Peroxide", and in addition to the normal manufacturer's pamphlets there is an extra sheet with "Instructions for Catalysing LR White Resin". Here the catalyst is described as "benzoyl peroxide powder catalyst". From information given by the manufacturer (but not found in their pamphlets/information sheet), the "benzoyl peroxide powder catalyst" is composed of 50% dibenzoyl peroxide and 50% unknown chemical stabiliser(s). No warnings as to the dangerous nature of the substance are given, as is the case with dibenzoyl peroxide supplied by other sources (e.g. Agar Scientific; Sect. 5.1), or whether the chemical stabiliser(s) reduce the risks. Following addition of the "benzoyl peroxide powder catalyst" to the supplied LR White, the resin mixture can be used in exactly the same way as detailed earlier in this chapter for the manufacturer's original LR White resin grades. The importance of this development for extending the versatility of LR White lies in the ability to now vary the cross-linked density of the resin, comparative to that already in effect with the Lowicryl resins (Sect. 5.3.2.5). For this reason, and also to ensure reproducible success with routine handling of the resin, details for preparing and using the 'uncatalysed' resin are given below.

A question that may arise is whether the manufacturer's "benzoyl peroxide powder catalyst" can be replaced by dibenzoyl peroxide (DBP) from another source.

In experiments from our laboratory no differences were observed between using the same amounts (i.e. the actual weight of DBP, see Sect. 5.1) of either compound on the polymerisation of 'uncatalysed' LR White by heat or chemical catalytic methods (J.A. Hobot, unpublished results). However, the 50% composition of unknown chemical stabiliser(s) may be important in infiltration or tissue penetrative properties of the LR White resin. For routine use of 'uncatalysed' LR White it is recommended that the manufacturer's powder catalyst is added, but this does not have to be carried out upon receiving the resin. There are two options: (i) Add all 4.5 g of the "benzoyl peroxide powder catalyst" to 500 g of 'uncatalysed' LR White resin, and mix by bubbling dry nitrogen through the mixture. (This ensures efficient dissolution of the catalyst and guards against the introduction of oxygen which is an inhibitor of acrylic resin polymerisation). *The resin is now stored at 4°C and used in exactly the same way as the methods detailed above for LR White heat and chemical catalytic polymerisation.* (ii) Add just the required amounts of the catalyst to aliquots of the resin for carrying out a particular embedding. The second option is recommended, as the shelf-life of the 'uncatalysed' LR White (stored at 4°C) is prolonged, and the possibility of carrying out UV-light polymerisation efficiently is maintained (Sect. 5.5.3).

The handling and storage of 'uncatalysed' LR White (hard grade) are exactly the same as the manufacturer's original LR White detailed above in this chapter and Chap. 6.

5.5.1 Heat Polymerisation Methods

For heat polymerisation at 50°C 0.9% by weight of "benzoyl peroxide powder catalyst" (actual weight of DBP is 0.45%) is added to 'uncatalysed' LR White monomer, and mixed by bubbling dry nitrogen through the mixture (10-15 min approximately). Polymerisation is for 24 hours in an aluminium block (Sect. 5.2.2.1)

5.5.2 Chemical Catalytic Polymerisation Methods

5.5.2.1 Room Temperature

0.9% by weight of the manufacturer's "benzoyl peroxide powder catalyst" (actual weight of DBP is 0.45%) is added to 'uncatalysed' LR White monomer at room temperature, and mixed by bubbling dry nitrogen through the mixture (10-15 min approximately). 1.5 µl of the manufacturer's accelerator are added per 1 ml of the now 'catalysed' resin, and stirred by using an orange stick for 30 sec. The remaining procedure is as per Sect. 5.3.1.1. The 'catalysed' resin (without accelerator) is used in all infiltration steps.

5.5.2.2 Cold Temperature (0°C)

0.9% by weight of the manufacturer's "benzoyl peroxide powder catalyst" (actual weight of DBP is 0.45%) is added to 'uncatalysed' LR White monomer at room temperature, and mixed by bubbling dry nitrogen through the mixture (10-15 min approximately). The mixture is cooled to 0°C and 1.5 µl of the manufacturer's accelerator are added per 1 ml of the now 'catalysed' resin, and stirred by using an orange stick for 30 sec. The remaining procedure is as per Sect. 5.3.1.2. The 'catalysed' resin (without accelerator) is used in all infiltration steps.

5.5.2.3 Cold Temperature (-20°C)

The procedure and precautions taken are the same as for 0°C above (Sect. 5.5.2.2), except that the amount of accelerator added can be increased up to 3 µl per 1 ml of resin. Polymerisation is at -20°C, and blocks can be stored at -40°C.

5.5.3 Ultraviolet Light Polymerisation Methods

For the manufacturer's original LR White resin no additions are necessary for UV-light polymerisation by an indirect source within 24 hours, as dibenzoyl peroxide is already present in the resin mixture (Sect. 5.4.2). 'Uncatalysed' LR White does not polymerise with UV-light unless the appropriate additions are made. Surprisingly, however, the addition of the manufacturer's "benzoyl peroxide powder catalyst" to 'uncatalysed' LR White does not result in polymerisation using an indirect UV-light source within 24 hours, but necessitates prolonging the polymerisation time to 48 hours. However, the blocks produced are not consistent with those obtainable under the standard protocols for acrylic resins (Sect. 5.4). Only approximately 5-10% of the resin block at the bottom of the gelatin capsule is hardened (J.A. Hobot, unpublished results). Using an alternative dibenzoyl peroxide from a different source (Agar Scientific) produced the same results. Switching to a direct UV-light source instead yielded polymerised blocks in both cases after 24 hours. However, the initial use of a direct UV-light source is not recommended, as this can lead to a too rapid polmerisation of the blocks with a consequent unacceptable temperature rise and shrinkage of the specimen (Carlemalm et al, 1982a; Sect. 1.3.4). To achieve UV-light polymerisation with an indirect source after 24 hours, dibenzoyl peroxide is omitted and the initiators benzoin ethyl ether or bezoin methyl ether are used instead. A further 48-72 hours polymerisation under a direct UV-light source follow to improve the sectioning qualities of the resin. The apparatus and methods for polymerisation are as described earlier (Sect. 5.4.1).

5.5.3.1 Room Temperature

For UV-light polymerisation of 'uncatalysed' LR White at room temperature 50 mg of the initiator, benzoin ethyl ether, are added per 10 g of resin. Handling and polymerisation conditions are as set out in Sect. 5.4.2.1.

5.5.3.2 Cold Temperatures (0°C to -20°C)

For UV-light polymerisation of 'uncatalysed' LR White at 0°C to -10°C, 50 mg of the initiator, benzoin ethyl ether, are added per 10 g of resin. For polymerisation at temperatures below -10°C and down to -20°C, 50 mg of benzoin methyl ether per 10 g of resin are added. Handling and polymerisation conditions are as set out in Sect. 5.4.2.2.

6 Handling Resin Blocks

6.1 Sectioning Blocks

6.1.1 Epoxy Resins

Epoxides have excellent sectioning properties. Glass or diamond knives can be used and for semithin sections the knives can be used dry. 0.5-1 µm sections can be picked up in the jaws of fine forceps and floated on to the surface of droplets of distilled water on glass-slides. They are dried down with gentle heat (< 60°C, 2-24 hours) and, without adhesive, will generally stick firmly to the surface of the glass. Problems with wrinkling or creasing can sometimes be alleviated by wafting chloroform on a cottonwool bud over the section while it is on the water droplet, prior to drying down. The hydrophobic surface of glass-slides "subbed" with chrome-alum gelatin (Pappas, 1971) is particularly useful for creating very rounded water droplets with a high meniscus. When warmed the reduction in surface tension is often strong enough to pull out wrinkles without recourse to organic solvents. Gelatin is digested by proteolytic enzymes and, therefore, should be avoided if section proteolysis for immunocytochemistry or *in situ* hybridisation is envisaged.

Thin sections, 30-100 nm in thickness, have to be floated out on water in a "knife-boat". Diamond knives come complete with boats attached. For glass-knives, boats can be preformed plastic reservoirs (Leica; the correct thickness of glass for these "boats" is ordered from Leica) attached to the knife with melted wax or made up for use from sticky tapes. Should wrinkling be a problem, acetone (30-70%) can be added to the water in the boat to reduce surface tension. Most modern ultratomes have variable possibilities for the speed of cut and the attack angle of the knife (these may vary between glass-knives of different break-angle and diamond-knives) and their makers provide detailed information on optimising settings for the commoner epoxy resins such as Araldite or Epon (Reid, 1975).

The thickness of sections can be judged from their interference colours on the surface of the water in the boat. Grey to white sections are thinnest at about 30-60 nm, pale straw darkening to gold are about 70-100 nm thick. In theory, the thinner the section the higher is its resolution. Sections are picked up, unsupported on the rough (matt) side of ethanol or acetone washed metal grids. 300 mesh, high transmission, nickel grids with hexagonal apertures are recommended for cytochemistry and immunocytochemistry. They give good section support leaving viewing minimally obstructed by grid-bars, and they are almost unreactive with immunoreagents even after prolonged exposure. Gold grids have no magnetic attraction for forceps like nickel grids, but are expensive.

Epoxy resins are hydrophobic and three-dimensionally bonded with enormous strength so that, even unsupported, thin sections are very stable in the electron beam.

6.1.2 Acrylic Resins

6.1.2.1 LR White (LR Gold)

Under-cured, hard-grade LR resin blocks section easily for almost any requirement. In general 350 nm has been found to be an economic thickness for semithin sections, being thick enough for most routine, cytochemical and immunocytochemical stains. They are, however, too thin for "dry" sectioning and should be cut on to water. Unlike epoxides the hydrophilic, acrylic block-face has an attraction for water and floods if the water level in the knife-boat is too high (i.e. as high as would be used for sectioning Araldite or Epon).

Large block faces may be cut but the sections will expand on the surface of the water in the knife-boat (depending on the degree of polymerisation) so some allowance for swelling must be made. 350 nm semithin sections are picked up on one of the jaws of a clean pair of fine forceps and transferred to a droplet of water on a glass slide. It is made much easier if the slides are "subbed" with chrome-alum gelatin (Pappas, 1971) which causes the water droplets to stand up higher. In addition, when the sections are placed in an accurately set oven at 50°C the increased surface tension pulls the sections flat during drying-down. If the water droplets are kept to 100-150 μl, the sections are ready for labelling or staining in 2-3 hours.

Most electron microscopical requirements are satisfied by 80-110 nm thick sections (gold - the refractive index of LR White is slightly less than that of epoxides), although much thinner ones can be cut. Very large block faces (1-3 mm) can be thin-sectioned. The conditions for sectioning, apart from observing a low meniscus in the knife-boat (see semithin sectioning above), are much the same as would be used for Epon 812 (Hayat, 1989) but the speed of cut should be faster than would normally be appropriate for Epon.

Glass or diamond knives can be used although it is easier to arrange the water level in the "boat" of a glass knife. Sections are placed unsupported on the smooth (shiny) side of ethanol (or acetone) washed metal grids. The sections are allowed to air-dry gently, held clamped in the ends of a pair of fine forceps. Excess liquid is removed from between the blades of the forceps with non-fibrous filter-paper (Whatman No. 50, hardened), but the grids must not be blotted or drained prior to air-drying or serious folding or tearing of sections can result.

When the hard-grade LR White resin is more fully cured it can be very hard and brittle and is often a darker shade of yellow than "correctly"-cured resin. Large block faces may be difficult to cut for semithin sections at thicknesses of more than 0.5 μm and have to be very much reduced in area for thin sectioning. Thin sections are quite stable in the electron beam, but the resin is less polar (more hydrophobic) and the tissue embedded in it much less sensitive to routine, cytochemical and immunocytochemical stains.

Care should be excersised when first examining under-polymerised LR White sections in the electron microscope. Polymerisation will be completed by the electron beam which should, therefore, be well spread at low magnification and high kV (80-120). After a period of a few seconds equilibration, the section can be taken up to much higher magnifications and greater beam current without detriment.

6.1.2.2 Lowicryls

Blocks should be trimmed to size for semithin and thin sectioning using a sharp razor-blade provided only empty resin is removed. To trim resin embedded tissue it is advisable to use a glass-knife in order to avoid vibrational damage to delicate tissue-resin interfaces. The method of sectioning is similar to that employed with LR White (Sect. 6.1.2.1). Lowicryls K4M and K11M are the most difficult resins to section. They are the most hydrophilic and require precise levels of water in the knife-boat to prevent it from flooding on to the block face. Large blocks can be sectioned with practise but the larger the block-face the lower should be the meniscus in the boat. The cutting speed is around 2-5 mm/sec and should be increased for bigger blocks. The less hydrophilic resins, Lowicryls HM20 and HM23, are easier to section. Glass or diamond knives are suitable.

As with other acrylics (LR resins above, Sect. 6.1.2.1), sections should be placed unsupported on the smooth or shiny side of ethanol or acetone washed nickel or gold grids for cytochemistry or immunocytochemistry. The sections are allowed to air-dry gently, held clamped in the ends of a pair of fine forceps. Excess liquid is removed from between the blades of the forceps with non-fibrous filter-paper (Whatman No. 50, hardened), but the grids must not be blotted or drained prior to air-drying or serious folding or tearing of sections can result. Copper grids often react with immunoreagents, although this can be avoided by dipping them in 0.05% collodion in amyl acetate. After draining and drying, they can be used like the other grids for mounting sections without support. Following immunolabelling and counterstaining, further support may be given by lightly carbon coating the section in a vacuum coating-unit. Alternatively, if required, grids can be coated with, for example, a collodion film from a 2% solution in amyl acetate and carbon-coated.

The same precautions as are used for viewing LR White sections in the electron microscope (Sect. 6.1.2.1) should be observed with Lowicryl sections.

6.2 Storing Blocks

6.2.1 Epoxy Resins

Epoxy resin blocks are very stable and need no special storage conditions. Blocks 30 years old and stored at room temperature are still sectioning well (G.R. Newman, personal observations). Loss of immunocytochemical reactivity with time has been impossible to guage.

6.2.2 Acrylic Resins

6.2.2.1 LR White (LR Gold)

Since chemically polymerised LR White blocks can be further polymerised at 50°C it seems likely that a slow polymerisation could continue in all blocks if left at ambient temperature, particularly in hot climates. In fact, blocks stored at room temperature for up to seven years have been shown to retain their immunoreactivity, but these were largely blocks which had been removed from their gelatin capsules and were, therefore, exposed to air. Oxygen is a powerful inhibitor of polymerisation of all acrylics and blocks which have been opened need very strenuous heating methods to further polymerise them. It is probably this factor which accounts for the apparent stability of stored blocks. One exception is seen in intracellular immunoglobulins such as those manufactured by and stored in plasma cells. A diminution of reactivity can be seen in, for example, formalin-fixed tonsil embedded in LR White in a relatively short period of time (2-3 months) if stored at room temperature. The exact reason for this is unknown, although, it is prevented by storing the blocks at temperatures below -20°C. To allow for this phenomenom being more widespread than has been suspected, particularly in sealed blocks, and to preserve the maximum immunoreactivity of blocks, it would seem a sensible policy to store at least some at low temperature in a deep-freeze. In this way continuity between experiments is assured. If it is not possible to store blocks at low temperatures, storage at room temperature can be conveniently done after the gelatin capsules have been removed. Exposure of the resin blocks to the atmosphere should allow for the inhibitory action of oxygen to prevent any futher polymerisation taking place.

6.2.2.2 Lowicryls

The only long-term storage problem with Lowicryl blocks is their tendency to absorb atmospheric water. It is advisable to seal them, especially if the capsules are opened, in an airtight container in the presence of a dessicant (silica gel) or to store them in a bench dessicator with dessicant. Following heat or chemical catalytic polymerisation, they may also be stored in a deep-freeze like LR White (Sect. 6.2.2.1), or have their gelatin capsules removed as dicussed above (Sect. 6.2.2.1).

PART II: ON-SECTION IMMUNOLABELLING

7 Strategies in Immunolabelling

7.1 Introduction

Albert Coons (Coons et al, 1941) began the science of immunocytochemistry when, in the light microscope, he demonstrated the possibility of localising tissue substances with fluorescent antibodies. Since then numerous ingenious markers have been attached to antisera and other macromolecules in order to identify tissue and cellular substances both in the light and the electron microscope. Ferritin (Singer, 1959) was the first EM marker, followed by enzymes, the most efficient of which for EM proved to be horse-radish peroxidase (Nakane and Pierce, 1966), because diaminobenzidine (DAB), one of its chromogens, could be made electron dense with osmium (Seligman et al, 1968). Peroxidase/DAB has unique properties making it a useful LM marker with EM potential. Colloidal gold was introduced as an EM immunocytochemical marker by Faulk and Taylor (1971), although its value as an electron dense tracer had been established by Feldherr and Marshall (1962). Its versatility and simplicity of manufacture and use rapidly established it as the EM marker of choice. Today, imunocolloidal gold and immunoperoxidase/DAB make up the bulk of immunolabelling methods and are summarised and compared in Fig. 6.

Colloidal gold, as evident in the wine red staining of early Chinese glass-ware, has been in use for thousands of years. However, much more recently the typically brilliant red suspensions of colloidal gold have been generated by the reduction of gold salts, usually chlorauric acid, in aqueous solution. Faraday (1857) established that the addition of electrolytes to colloidal gold suspensions changed their colour from red to blue as the stability of the colloid was destroyed and the particles flocculated. Further, he showed that the addition of protein inhibited this effect. Faraday's observations form the basis of modern colloidal gold preparative methodology which has resulted in the production of a variety of very characteristic, electron dense markers of great value in cytochemical and immunocytochemical electron and light microscopy (Handley, 1989a, 1989b). The size of the gold particles can be very accurately controlled by varying the method of gold salt reduction so that sols with homogenous particle sizes from 1 nm to 150 nm can be reproducibly prepared, often with coefficients of variation of less than 10%.

The ability of colloidal gold to adsorb substances on to the surface of its particles makes it invaluable as a cytochemical and immunocytochemical marker. Antibodies and many other macromolecules can be attached to colloidal gold without seriously damaging their biological reactivity. The macromolecular coating is thought to be bound to the gold by powerful electrostatic forces and has the effect of stabilising the colloidal state of the sol, though this area is little understood.

The particulate nature of colloidal gold, so easy for counting, has made it very tempting for quantitation (Griffiths and Hoppeler, 1986; Lucocq, 1992; Posthuma et al, 1987). However, the number of immunoglobulin molecules or other macromolecules adsorbed on to the surface of a colloidal gold particle is thought to vary with the size of the particle. Various estimates have been voiced based on the quantity of protein needed to stabilise a given sol (Geoghagen, 1991; Horisberger and Clerc, 1985), but it has been impossible to actually measure the relationship of individual gold particles to adsorbed macromolecules on their surface. This relationship is important in establishing the ratio of colloidal gold particles to antigenic sites in, for example, an immunocytochemical localisation. The number of colloidal gold particles marking a given site increases as (a) the particle size decreases (Slot and Geuze, 1981), or (b) the number of macromolecules (e.g. antibodies, protein A, etc.) adsorbed to the colloidal gold particle increases (Ghitescu and Bendayan, 1990). When added to the complex problems of preserving tissue antigens prior to localisation it makes absolute estimates of amounts of the antigen in most cases too hypothetical to be of any practical value. Nonetheless, provided scrupulous attention is paid to uniformity and reproducibility of method, with the inclusion of appropriate controls (Sect. 8.3), comparative studies are perfectly possible.

There are many descriptions for the manufacture of the various forms of colloidal gold sols and their conjugation to different macromolecules and this will not be dealt with here (Faulk and Taylor, 1971; Handley, 1989b; Roth, 1983). Most of those needed for cytochemistry and immunocytochemistry are commercially available, typically in 1, 3, 5, 10, 15, 20 and 40 nm sizes.

The enzyme peroxidase was one of the earliest markers to be adapted to electron microscopy (Nakane and Pierce, 1966). Originally it was extracted from plants, being heat stable and easily purified. Unfortunately, in addition to peroxidase and catalase (which reacts with the same substrate) there are numerous pseudoperoxidatic agencies that occur in many animal tissues, posing the problem of their abolition (Sect. 8.1.3). The reaction of peroxidase with its substrate, hydrogen peroxide, involves an exchange of electrons which can be accompanied by the precipitation of a suitable chromogen. One such is diaminobenzidine (DAB), which polymerises and precipitates when it donates electrons to the peroxidase/hydrogen peroxide complex (Seligman et al, 1968), generating a brown insoluble colour product easily seen in the light microscope but only slightly electron dense. DAB reacts strongly with osmium tetroxide (osmium black - Seligman et al, 1968) which provides added electron density for viewing DAB in the electron microscope.

However, sodium gold chloride, a much less hazardous alternative, will also markedly increase the electron density of DAB. It is specific for DAB and does not increase the electron density of background tissue (Newman et al, 1983b, 1983c). The high affinity of DAB for sodium gold chloride renders it significantly more electron dense than with osmium (Sect. 8.2.5) and provides a means for the photointensification of DAB for light and electron microscopy (Newman et al, 1983b; Sect. 8.2.6).

DAB still has a significant part to play in modern immunoelectron microscopy providing "panoramic" views of antibody reactive sites at low magnification, as opposed to immunocolloidal gold which can be very difficult to see (Newman and Jasani, 1984b). The strikingly different EM appearances of colloidal gold and DAB/gold chloride enable them to be used together on the same section for EM double immunolabelling (Sect. 7.4).

The immunocytochemical techniques mentioned above will have an end-result whose sensitivity will depend not only upon the correct application of the appropriate labelling or staining method, but also upon the antigenic reactivity retained within the tissue. The latter has been the theme of the first part of this book, and the strategic choices made there will now affect the outcome with the techniques discussed in the book's second part.

Colloidal gold, and subsequent silver intensification methods if required, are applicable to all resins. DAB/peroxidase techniques will not be successful with Lowicryls K4M and K11M, but will be possible with Lowicryls HM20 and HM23. The more polar Lowicryl resins are very reactive with basic stains and dyes, including DAB, and therefore give an unacceptably high background (J.A. Hobot and G.R. Newman, unpublished results). Thus, the sensitive processing protocols at room temperature involving low fixative concentrations and partial dehydration (Sect. 3.3.3), whilst fine for Lowicryls K4M and K11M when using colloidal gold, will have to be used with LR White if peroxidase techniques are planned. Naturally, low fixative concentrations and PLT can be employed if Lowicryls HM20 or HM23 are wanted for peroxidase techniques. The same reasoning will apply for double immunolabelling techniques exploiting colloidal gold/peroxidase (Sects. 7.4/9.4.2).

7.2 Colloidal Gold Strategies

7.2.1 Direct Methods

The primary reagent might be an antibody, a lectin, an enzyme or some sort of ligand linked directly to colloidal gold. Early direct methods empoying various antisera are now rarely used, although lectins conjugated to colloidal gold have been used on LR White (Ellinger and Pavelka, 1985) and Lowicryl K4M (Roth, 1983) sections. Enzymes have also been conjugated directly to colloidal gold and are used to localise their substrates (Bendayan, 1981; Londoño et al, 1989). An advantage of the direct method, apart from its brevity, is that, theoretically, multi-immunolabelling is possible for the EM by having the various primary reagents coupled to different sizes of colloidal gold particle. This system is no good for LM in which the colloidal gold is visualised by photochemistry (7.2.4). In practice, the amount of work required to produce the conjugates is daunting, the concentrations of colloidal gold conjugate needed for a response can be high, which is expensive and wasteful, and the levels of labelling, particularly with the larger gold sizes, are often very poor.

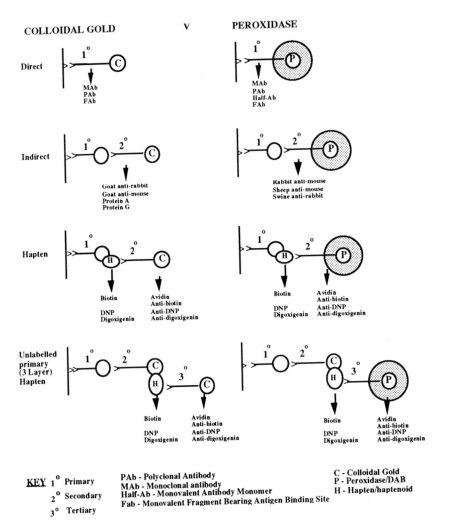

Figure 6. On-Section Immunolabelling Methods

7.2.2 Indirect Methods

Direct methods have largely given way to much more sensitive, two step, indirect techniques in which the primary reagent of the first step is localised by a colloidal gold-linked detection system in the second step.

7.2.2.1 Immunogold Staining (IGS) or Labelling

In IGS the colloidal gold is linked to an antibody which recognises natural determinants on the primary reagent for its localisation. The primary reagent can be an antibody, a lectin or a ligand such as a hormone, and the secondary detection system is based on specific antibodies raised against them which are coupled to the colloidal gold (De Mey, 1983). Fortunately, most antibodies are raised in rabbits (polyclonal) or mice (monoclonal) so that antibodies specific simply to the IgG of these two species each linked to colloidal gold, will suffice for the majority of immunolocalisation studies. Nonetheless, for each primary reagent other than rabbit or mouse IgG, specific antibodies have to be raised, or bought, and coupled to colloidal gold. In addition, non-specific labelling occurs when the natural determinants of the primary reagent, recognised by the secondary detection system, are also found in the tissue.

7.2.2.2 Protein A-Gold

A lowering of background may be obtained with Protein A conjugated to colloidal gold. Protein A is an extract from the bacterium *Staphylococcus aureus* which recognises and attaches to the Fc groups of some antibodies (Romano and Romano, 1977; Roth et al, 1978). When used as a secondary detection system, it ensures that only the primary antiserum is localised. Unfortunately, Protein A can only localise to immunoglobulins and then only to some classes (Björck and Kronvall, 1984). The mouse monoclonal antibody is one notable class for which Protein A often has a very poor affinity. When using Protein A to localise monoclonal antibodies a polyclonal antimouse monoclonal bridge antibody is often necessary. Lectins and ligands will also need polyclonal bridge antibodies before the colloidal gold Protein A complex can localise them. Obviously some of the value of using Protein A is lost in these situations.

A novel use of protein A has been to avoid its interaction with colloidal gold and instead to prepare solutions of Boronated Protein A. The protein A is still reactive with Fc groups on the antibody, but the antibody is now visualised as a boron signal by Electron Spectroscopic Imaging, available as standard with a Zeiss EM 902 (Bendayan et al, 1989). The location of the antibody can be observed directly on the tissue section, in much the same way as colloidal gold labelling in transmission electron microscopy. The spatial resolution is supposedly improved in terms of the distance the antibody may be detected from its specific antigenic site, and would depend on the amount of boronated probe (i.e. the signal strength) attached to the Fc groups. With colloidal gold, the spatial resolution depends upon its size (diameter) and the position of attachment of the colloidal gold particle to the Fc group, and there will most likely be more than one gold particle attached to the Fc group. However, with small gold particles (5nm), this improvement could be limited to less than 5nm (i.e. the gold particle diameter), bearing in mind that the length of an IgG molecule is 10-22nm depending upon its configuration. The use of small colloidal gold particles attached to Fab fragments can improve the spatial resolution (Baschong and Wrigley, 1990). Even so, steric hindrance caused by the antibody molecules themselves can also be a limiting factor in both boronated probe and colloidal gold marker systems. Finally, the first results presented by Bendayan et al (1989) with the boronated probe marker system resemble pictures that can be obtained using peroxidase/DAB.

7.2.2.3   *Protein G-Gold*

Another class of bacterial protein for localising antibodies, protein G, has been isolated by Björck and Kronvall (1984) from a Streptococcal strain. Its advantage may be that it recognises a broader range of immunoglobulins than protein A (Björck and Kronvall, 1984), although whether this includes important classes such as the mouse monoclonals is under dispute (Bendayan, 1987; Bendayan and Garzon, 1988; Taatjes et al, 1987).

7.2.2.4   *Protein AG-Gold*

Another addition to the group of bacterial proteins that bind to colloidal gold has been a chimeric Protein AG, which has been engineered from Staphylococcal Protein A and Streptococcal Protein G (Eliasson et al, 1988). Protein AG combines the properties of both Protein A and Protein G, and shows in some cases a higher avidity for classes of human IgG (Eliasson et al, 1989). It therefore may offer a more versatile alternative to its parental proteins. Its application to immunolabelling of resin sections is no different from the other two proteins (Ghitescu et al, 1991).

Figure 7. Kidney tubule of rat prepared as described in fig. 4. Biotinylated e-pha (erythrophytohaemagglutinin - *Phaeseolus vulgaris*; Sigma) labelling. Lectin localised with (a) Avidin conjugate to 15 nm colloidal gold particles. (b) Antibiotin antibody conjugated to 15 nm colloidal gold particles (see also Fig. 9). (b) probably reflects the true level of labelling since colloidal gold often obstructs the biotin 'docking ports' when conjugated to avidin, resulting in low labelling efficiency. (Mag. for both x 30,000).

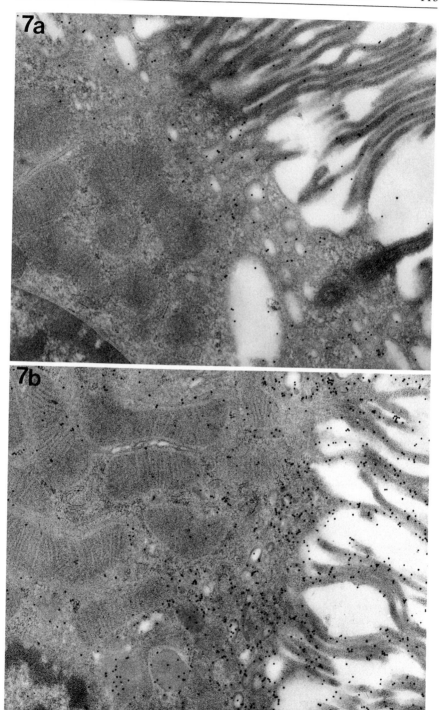

7.2.3   Hapten- (and Haptenoid-) Based Methods

In the IGS and Protein A-, G-, AG-gold methods, the gold conjugate recognises natural determinants on the primary reagent. These determinants, or something very similar, may be present elsewhere in the tissue and this often gives rise to high background. More seriously, IGS methods can never be used on tissue homologous with the gold conjugate. For example antimouse immunoglobulin/colloidal gold cannot be used on mouse tissues without generating prohibitive background staining. Many valuable, naturally occuring, clinical antisera cannot be used as immunocytochemical probes on human tissue for the same reason. In addition, Protein A and G do not recognise some classes of commonly employed antisera. In fact, in order to be able to cope with all eventualities, a vast array of different antibody conjugates would be necessary.

These problems are solved by chemically labelling the primary antibody with an artificial substance, or hapten, and localising it with a secondary detection system aimed specifically at the hapten. In addition, both the specificity of the localisation and the generality of the method can be improved. Now the colloidal gold linked secondary detection system is rationalised for all primary reagents, being designed to locate only to the hapten.

To create such a system raises many problems. The labelling method must be simple and reproducible. It must not be deleterious to the primary reagent which should retain its biological activity. The hapten should be easily recognisable to a suitable detection system and should not occur naturally in the tissue under study. Many systems have been written about but only the two most practical, (biotin and DNP), will be described here.

Haptenoids act as haptens but are natural products often borrowed from other applications such as FITC or digoxigenin. The best known haptenoid is biotin.

Figure 9. Semithin section of kidney tubule prepared as in Figs. 4 and 7. Biotinylated e-pha lectin labelling with colloidal gold. The colloidal gold has been visualised by silver intensification (Sect. 7.2.4). The basement membrane, microvillus border and lysosomes of proximal tubules are stained. The distal tubules remain unstained. (Mag. x 400).

Figure 10. Semithin section of exocrine pancreas of rat, perfusion-fixed with 1% glutaraldehyde and embedded in LR White following partial dehydration and cold chemical catalytic polymerisation. Immunocolloidal gold silver staining (IGSS). A polyclonal antiserum against rat anionic trypsin has been localised by goat anti-rabbit/colloidal gold to the storage granules of the acinar cells. Trypsin released into ducts has also been localised (arrows). The colloidal gold has been visualised by silver intensification. (Mag. x 250).

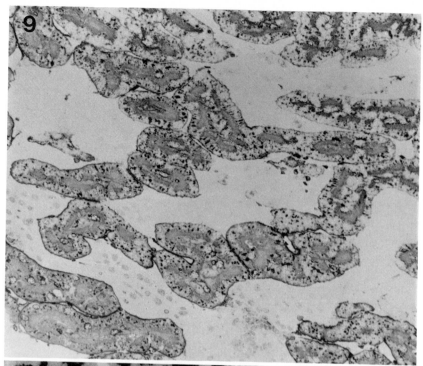

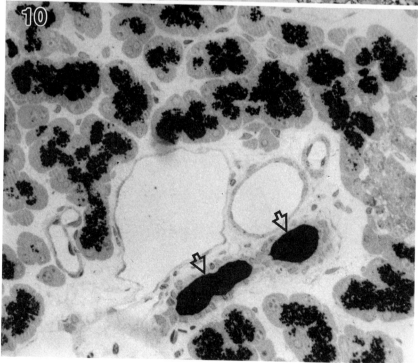

7.2.3.1 Biotin

This is probably the most extensively used haptenoid. It is an extract of egg-yolk, and biotinylated primary reagents are now widely available. Many localisation systems have been designed around it. Most of these depend on the very high affinity of avidin for biotin (even higher than that of polyclonal antibodies for their antigens). Avidin is an extract of egg-white which unfortunately also has an affinity for nuclear protein. This can be overcome by using streptavidin, a more expensive extract from Streptococcal bacteria, which has become popular, particularly for studies on nuclear immunocytochemistry and *in situ* hybridisation. Some avidin/biotin systems employ a biotinylated secondary antibody so that they can be used with unlabelled primary reagents. The labelled secondary antibody acts as an intermediary between the unlabelled primary reagent and the remainder of the detection method. The second step is therefore, once again, dependant on natural determinants thus defeating some of the advantages of the hapten method.

Two possibilities exist for the localisation of biotinylated primary reagents (Bonnard et al, 1984). The most common method employs avidin. Avidin and streptavidin will complex with colloidal gold, but in doing so it is believed that some or all of its four "docking ports" may be obstructed and prevent its binding optimally to biotin. The second approach, based on a high affinity, very specific antibody to biotin, solves this problem. Endogenous biotin or biotin-like substances in tissue can give rise to problems of non-specific labelling.

7.2.3.2 Dinitrophenyl (DNP)

This hapten has been designed for easy, non-deleterious labelling (Jasani and Williams, 1980; Jasani et al, 1981; Hewlins et al, 1984). Almost any primary reagent can be labelled with DNP which does not occur naturally so there are no problems of its ocurring endogenously. The compound is bright yellow and covalently binds to substances when they are exposed to it, in solution, for 2 to 4 hours. These, too, become bright yellow in colour and are easily separated from excess of the hapten by passing the mixtures down a minicolumn (Hewlins et al, 1984). DNP-labelled materials act in exactly the same way as their unlabelled counterparts, retaining their biological activity and still being recognisable to the same antisera. An antibody to DNP conjugated to colloidal gold is used to localise the hapten. DNP labelled primary antisera are still localisable by the IGS method and by Protein A-, G- or AG-gold.

7.2.4 Silver Intensification

All forms of colloidal gold are difficult to see in the light microscope and 1 nm colloidal gold particles are often difficult to detect in the electron microscope. However, no complex or expensive additions to the standard light or electron microscope are required to visualise colloidal gold deposits. A sensitive, simple one step photochemical procedure can be conducted in daylight under normal laboratory conditions using simple laboratory utensils (C-Gold; BioClinical Services) to enlarge and visualise colloidal gold deposits (Fig. 8).

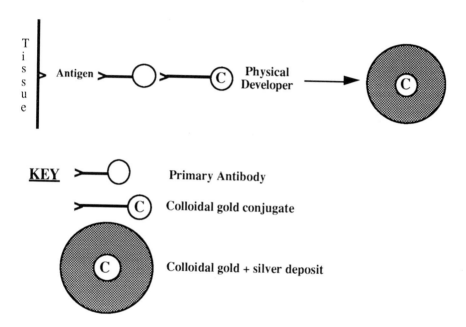

Figure 8. Immuno-Gold Silver Staining (IGSS)

7.3 Peroxidase Strategies

7.3.1 Direct Methods

Superficially, a single step method sounds very convenient but the complexity of having to covalently bind peroxidase to each individual primary antibody and the lack of sensitivity when they are applied have made these methods very unpopular. Peroxidase has been directly linked to hemi-antibody fractions and Fab with a view to higher sensitivity, stoichiometric immunolabelling (Fujiwara et al, 1981; Yoshitake et al, 1982).

7.3.2 Indirect Methods

The two step indirect method localises the primary reagent, in the second step, with an immunoglobulin covalently linked to peroxidase. Improvements in antibody production and the chemistry for conjugating them to viable enzyme markers, have opened the way for more accurate labelling of sections with peroxidase/DAB using these secondary detection methods. The conjugate has a high molecular weight but is much smaller than the PAP complex (Sect. 7.3.3). In theory there is at least one peroxidase molecule per antibody molecule (PAP has a 3:2 ratio) and probably much more than this in practice. This improved ratio coupled with the use of sodium gold chloride ensures denser, better localised DAB deposits than are achievable with PAP.

7.3.3 The Peroxidase Antiperoxidase (PAP) Method

The sensitive but complicated, three step, bridge technique of Sternberger (1979), which employs preformed complexes of antiperoxidase antibody with peroxidase enzyme (PAP) in the third step, was once favoured for "on-grid" immunolabelling of epoxy resin sections (Pelletier and Morell, 1984). Semithin sections, etched with sodium ethoxide (Sect. 8.1.1.1) frequently give staining intensities with the PAP technique which are as good as those obtainable with rehydrated paraffin sections. When viewed in the electron microscope, however, PAP complexes are large enough for their characteristic ring-structure to be readily resolved. In fact, by far the greater part of the PAP complex consists of the immunoglobulin molecules of which there are thought to be three to every two much smaller peroxidase molecules. Following osmium treatment both the immunoglobulin fraction and the precipitated DAB are made more electron dense so that immunolabelling is seen as a rather untidy, often ill-defined particulate scattering, frequently obscuring the ultrastructural detail. Sodium gold chloride combines only with the DAB (Newman et al, 1983b; Sect. 8.2.5) giving more accurate localisation. Unfortunately, smaller amounts of DAB, relative to immunoglobulin, are deposited so that low contrast may result (Newman et al, 1983a; and unpublished results). In addition the PAP method suffers from the Bigbee effect (Bigbee et al, 1977) in which the bridge antibody becomes exhausted by excessive concentration of the primary reagent, giving a false negativity. This places a heavy reliance on dilution series to ensure nothing is missed.

   Immunoperoxidase methods which employ reagents with relatively small, monodispersed molecular structures have been shown to give higher densities of labelling particularly in sections of acrylic resins such as LR White (Newman and Jasani, 1984a)

7.3.4 Hapten- (and Haptenoid-) Based Methods

7.3.4.1 Biotin

The simplest method localises the biotin-labelled primary reagent with peroxidase-conjugated avidin and has been used for localising lectins in LR White embedded tissue (Jones and Stoddart, 1986). As with any other peroxidase system, DAB is a very suitable chromogen for EM viewing. The more complicated ABC method (Avidin Biotin Complex), in which a preformed complex of avidin and biotin/peroxidase is employed to localise the biotin-labelled primary reagent (Hsu et al, 1981), has been successfully applied to epoxy resin sections (Childs and Unabia, 1982). It is not recommended for use with acrylic sections because, like Sternberger's PAP, the very large complexes do not penetrate well and can give rise to less sensitive, untidy, surface staining. Methods which employ avidin or streptavidin as a bridge in a three layer sandwich method i.e. biotinylated primary reagent, followed by excess avidin, followed by biotin/peroxidase, (Bonnard et al, 1984) are potentially the most sensitive but are not popular. Like PAP they are complicated by the Bigbee phenomenon (Sect. 7.3.3). Avidin can be replaced by an antibody to biotin conjugated to peroxidase and used in much the same way as avidin/peroxidase. Antibiotin has little advantage over avidin when conjugated to peroxidase, although it may be important for conjugation to colloidal gold. Wide use of avidin/biotin has shown that biotin does occur in some tissue, notably liver and kidney, and can be the cause of unwanted background staining. It is possible but difficult to inhibit endogenous biotin and it may be simpler to switch to a hapten that has no naturally occuring equivalent.

7.3.4.2 Dinitrophenyl (DNP)

The usual method of localising DNP-labelled substances is with a very sensitive, three layer method, (DNP Hapten Sandwich Staining, DHSS), which has been used to localise TSH receptors in LR White embedded thyroid (Newman and Jasani, 1984a). The DNP-labelled substance is followed by a monoclonal anti-DNP bridge antiserum and this is followed by a DNP-labelled peroxidase enzyme marker (see Fig. 15 for an example of this method of labelling). DAB/sodium gold chloride is used as the chromogen for EM viewing. A four-layer system in which an unlabelled substance is localised with a DNP-labelled species specific secondary followed by anti-DNP bridge antiserum and DNP-labelled peroxidase, is even more sensitive, but, as with biotin, the system is lengthier and loses some of the advantages of primary hapten labelling. Both methods are complicated by the Bigbee effect (Sect. 7.3.3). A less sensitive indirect method utilises an antiserum to DNP conjugated to peroxidase (Dako). This method is shorter, more convenient and not prone to the Bigbee phenomenon.

7.3.5 Silver Intensification

DAB may not be visible in resin semithin sections, but, following slight modification, like colloidal gold, it can be photochemically amplified (Fig. 11), (DAB Silver Intensification Kit, Amersham International).

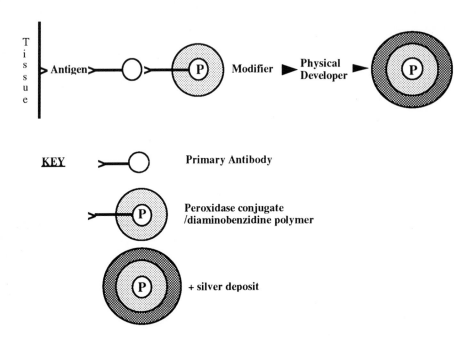

Figure 11. Immuno-Peroxidase Silver Staining (IPSS)

Figure 12. Formalin fixed decalcified bone embedded in LR White by the cold chemical catalytic method following partial dehydration. IGSS shows the outer membranes of osteocytes after the use of an antiserum to proteoglycan. Lightly counterstained in toluidine blue. Picture courtesy of Dr. Andrew Smith, School of Dentistry, University of Wales Colege of Medicine. (Mag. x 450).

Figure 13. 350 nm semithin section of pituitary of rat fixed by perfusion with 1% glutaraldehyde and embedded by the rapid room temperature method in LR White (Sect. 3.3.2). The DHSS immunoperoxidase method (Sect. 7.3.4) has been used following incubation of the sections in DNP-labelled anti-TSH antiserum. The DAB deposits were completely invisible before silver intensification. No counterstain. The section is very thin so that in places only part profiles of cells are visible. (Mag x 350).

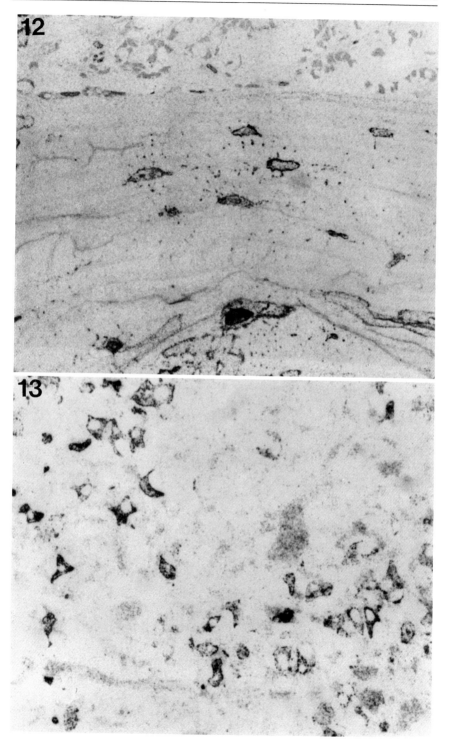

1. Sequential

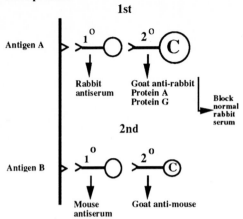

Antigen A is localised first with one size of gold particle. Spare valencies on the secondary detection system are blocked and antigen B is then localised with a different size of gold particle. The animal species of the primary antisera must not be cross-reactive. Steric hindrance may be a problem whether the smaller or the larger particle is used first.

2. Simultaneous

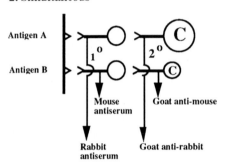

Both primary antisera are applied in a mixture, followed by a mixture of both secondary detection systems. Provided there is no cross-reaction this method can be used with antisera from any two different species using two sizes of gold particle and reduces the steric hindrance often shown by 1.

3. Hapten (sequential)

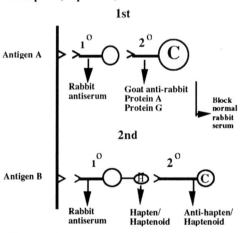

Different antisera from the same species can be localised provided one of them is labelled with a hapten or haptenoid. The staining must be sequential as in 1, or the secondary detection system for antigen A will localise antigen B inspite of the hapten/haptenoid label.

Figure 14. EM double imunolabelling using different sizes of colloidal gold particle.

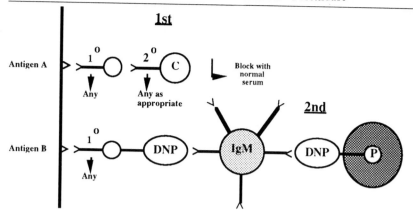

The method is sequential. Antigen A must be localised with immunocolloidal gold first. Antigen B is then localised by the immunoperoxidase, dinitrophenyl hapten sandwich staining (DHSS) method applied to a DNP-labelled primary antiserum. The antisera to A and B can be derived from the same species, but note that if the DHSS method is used first the secondary detection system used for localising anti-A may detect anti-B even though anti-B is DNP-labelled.

Alternatively, the primary antiserum against antigen B may be biotin labelled.

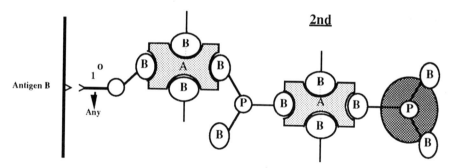

The method is similar. Antigen A is localised with immunocolloidal gold, the biotin labelled antiserum to antigen B can be localised with the avidin/biotin/peroxidase complex (ABC) as shown, with avidin used as a bridge or with an antibiotin peroxidase conjugate.

KEY

Dinitrophenyl   (DNP)   AntiDNP   (IgM)   Peroxidase/DAB   (P)
                                                (gold chloride for EM)

Biotin   (B)   Avidin   [A]

Figure 15.   EM double immunolabelling. Immunocolloidal gold combined
             with immunoperoxidase

7.4   EM Double Immunolabelling

LM double immunolabelling requires that different antigens are shown in different colours. The photointensification of DAB and all sizes of colloidal gold results in the same black staining which prohibits their use for LM double immunolabelling. However, differences in the electron microscopic appearances of different sizes of colloidal gold particles and between DAB and colloidal gold can be exploited for EM double immunolabelling.

Colloidal gold particles are very valuable for double immunolabelling because they can be manufactured in a variety of sizes, each size having a very small coefficient of variation. Thus two antigens can be differentiated by their localisation to gold particles of distinctly different sizes (Fig. 14). In its simplest form, double immunolabelling can occur if an immunocolloidal gold procedure using one size of gold particle is followed sequentially by another using a second size of colloidal gold particle (Fig. 14, diagram 1). The gold particles should be sufficiently different in size to be easily distinguished, say 5 nm and 15 nm, or 10 nm and 20 nm. To prevent unavoidable cross-reaction, the primary antisera of the two procedures must be from separate species. Spare valencies on the colloidal gold used for the first localisation can be blocked before commencement of the second incubation. One conspicuous disadvantage with this method is that sites common to both primary antisera will be occupied by the first and, therefore, blocked from the second (steric hindrance).

Figure 16. 1% glutaraldehyde immersion-fixed A-cell in the pancreas of rat embedded in LR White following partial dehydration and cold chemical catalytic polymerisation. Double immunolabelled with rabbit polyclonal antisera to glicentin and glucagon. The zoned alpha-granules are labelled with 20 nm gold particles largely in their peripheral regions, showing glicentin reactivity, and 10 nm gold particles in their central areas, showing glucagon reactivity. The anti-glicentin antiserum was used in conjunction with goat anti-rabbit antiserum conjugated to 20 nm colloidal gold. The anti-glucagon was DNP-labelled and shown second with an antiserum to DNP conjugated to 10 nm colloidal gold. Lightly counterstained in uranyl acetate and lead citrate. (Mag x 60,000).

Figure 17. BGPA fixed, surgically removed human pituitary embedded in LR White following partial dehydration and heat polymerisation. Double immunolabelled with colloidal gold and peroxidase/DAB. Rabbit anti-prolactin was used first in conjunction with colloidal gold. DNP-labelled anti-ACTH was used in conjunction with DHSS peroxidase DAB/gold chloride. Some granules are labelled with both colloidal gold and DAB/gold chloride. See also Fig. 23. (Mag. x 18,000).

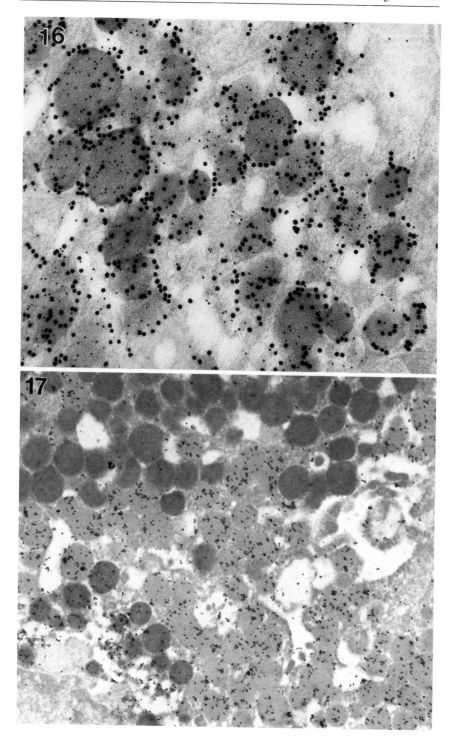

Two primary antisera can be applied simultaneously as a mixture (Fig. 14, diagram 2). The antisera must be from different animal species which do not cross-react. The secondary detection systems required to localise each primary antiserum must also be free from cross-reaction, either to the other primary antiserum or to each other. In theory, since both primary antisera are applied together, one does not block the localisation of the other, but, there may be a distinct difference in affinity between them. The antiserum with the higher affininty would then compete against the antiserum with the lower affinity for sites common to them both. The dilution of each primary antiserum must be arranged to establish the right balance.

Neither of the above methods copes with the localisation of primary antisera from the same species. For example, two polyclonal primary reagents, rabbit anti-A (A) and rabbit anti-B (B) would have to be localised sequentially. (A) could be localised first with goat anti-rabbit antibody linked, say, to one size of colloidal gold (GAR5) and, theoretically, (B) could be localised second with goat anti-rabbit antibody linked to another size of colloidal gold (GAR10). However, after the first procedure using (A) and GAR5 there would be spare valencies on both these antibodies leading to cross-reaction with the antisera of the second procedure and a confusion of gold particles of both sizes. (A) could be blocked with unconjugated GAR to prevent it reacting with GAR10 as well as GAR5, and the spare valencies on the GAR5 could be blocked with normal rabbit serum (NRS). Unfortunately, however, primary antiserum B would then cross-react with the unconjugated GAR on A and the GAR10 would localise the NRS used to block GAR5. In practice, blocking is rarely complete and very complex solutions would be needed to solve these problems. The same problems occur when Protein A (or G, AG) linked to colloidal gold is employed as the secondary detection system or if the systems are mixed.

The solution has been the development of hapten-based secondary detection systems. The localisation of two primary reagents from the same species on the same section can be acheived sequentially by labelling one of the antisera with a hapten (Fig. 14, diagram 3). If both primary antisera are, for example, rabbit polyclonals, immunocolloidal gold is used first to localise the unlabelled, primary antiserum. Unoccupied valencies on this antiserum or protein A/G/AG are blocked with NRS or rabbit IgG (RI) and the second, hapten-labelled, primary antibody is applied. The hapten can now be localised with anti-hapten antiserum linked to a second colloidal gold particle of different size (Newman, 1989).

When immunolabelling LR White sections, the combination of immunocolloidal gold followed by immunoperoxidase avoids the often serious problem of steric hindrance that occurs with two gold labels (Newman and Jasani, 1984b; Newman et al, 1986; Newman et al, 1989). This is because the colloidal gold and DAB occupy different sites in the section (Newman and Hobot, 1987). It is not advisable to use DAB with Lowicryls K4M or K11M (Sect. 7.1). Once again a sequential method must be used to avoid cross-reaction (Fig. 15), and the peroxidase method can be based on DNP- or biotin-labelled primary reagents.

In the most sophisticated use of haptens, simultaneous localisation of two antisera from the same species becomes possible. The first primary antiserum can be labelled with one hapten and the second with another. They can now be mixed and applied together, when they will compete for the same sites equally, thus reducing steric hindrance. A mixture of anti-hapten antibodies (or avidin) linked to different sizes of colloidal gold can localise both primary antisera simultaneously in the second step. Further, three, or possibly more, antigens from the same species may be localised by combining sequential and simultaneous methods. For example, one antigen could be localised with an unlabelled primary antiserum and immunocolloidal gold, followed by two, simultaneously localised as above. Three contrasting sizes of colloidal gold particle would be necessary and analysis of their distribution could be very complex. Presumably immunoperoxidase could be used to localise a fourth. These are largely untried, theoretical possibilities and protocols for them are not included here.

8   General Considerations

The procedures required to maximise the immunolabelling of tissue sections can be placed in to two groups, those which prepare the sections for staining, collectively called the section pretreatment, and the staining procedures themselves. The requirement for section pretreatment depends on the extent of fixation and the type of resin, with epoxy resin sections of post-osmicated tissue needing the most attention, and acrylic resin sections of aldehyde fixed tissue the least. Immunolabelling levels will be limited by the availability of antigen in the section but can be optimised. Much of this optimisation is empirical but some consideration of the factors involved can help to reduce the number of experiments.

Immunolabelling of resin sections is carried out after the sections (semi- or thin) have been mounted on glass slides or metallic grids as described in Chap. 6. Thin sections can be immunolabelled either by floating the grids on reagent droplets, or by immersing the grids fully in the reagent. The latter method can only be used for thin sections mounted on uncoated (naked) grids. In practice, the procedures employing immersion have been found to give a better response in terms of reproducible on-section immunolabelling or labelling, together with lower background and cleaner final preparations, than the flotation method (G.R. Newman and J.A. Hobot, unpublished results; immersion methods have been used mainly in the micrographs presented in this book and relevant publications of the authors). With the immersion method it is important to reduce the amount of carry-over from one solution to the next. It is simply accomplished by gently shaking off the excess between each transfer, but never allowing the grid to dry out. When using coated grids, care must be taken not to have reagent going on to the back of the grid, especially any colloidal gold solutions, as these have a preference to adhere to the film.

8.1   Resin Section Pretreatment

8.1.1   Etching (Epoxy Resin Sections only)

Removal of the resin, or etching, is unnecessary and impracticable with sections of acrylic resins. However, to promote the immunolabelling and counterstaining of epoxy resin sections, it may be essential. Unlike epoxy resins, acrylics do not react with tissue, but bond through it. The bonds linking an epoxide to tissue can be broken and the resin removed by etching. If the bonds of a cross-linked acrylic (e.g. LR White, LR Gold and the Lowicryls) are broken the tissue will come away with the resin, making etching largely ineffectual and possibly seriously damaging. Sometimes a spurious improvement is seen in immunolabelling following "light etching" of an acrylic resin section with organic solvent but this is due simply to the increased surface area (Sect. 10.3). The extreme hydrophobia of all epoxy resins has two serious consequences for cytochemistry and immunocytochemistry. It is almost impossible for aqueous solutions to penetrate sections at neutral pH, resulting in weak or non-existent reactions, and the resins have a strong attraction for hydrophilic reagents, such as those used in cytochemical and immunocytochemical reactions, sometimes leading to high nonspecific background staining. Etching may counteract these problems, but it is a subjective measure which can introduce problems of its own.

8.1.1.1 Semithin Sections

Epoxy resin can be removed from sections by immersing them in concentrated sodium ethoxide or methoxide. To prepare these etching solutions 100% ethanol or methanol is saturated with sodium hydroxide by standing it over hydroxide pellets for at least a week. The solution should be gently mixed, daily, by inversion. It becomes viscous and gradually goes brown, darkening as it ages.

The extent of etching required is very variable from epoxide to epoxide and even from batch to batch of the same epoxide. It is further complicated by the way in which sodium ethoxide and methoxide mature on keeping, becoming more effective at removing resin as they get older. Inevitably some antigens will be damaged or extracted by a very active, concentrated etching solution. In order, therefore, maximally to preserve antigens the resin should, if possible, only be partially removed by brief (1-5 minutes) exposure of the section to the etching agent (or a dilution of it). Inevitably the requirement for etching has to be empirical and for some antigens the resin may have to be completely removed.

The etching solution should be washed away thoroughly with several changes of distilled water.

Figure 18. Part of a glomerulus from a kidney biopsy immersion-fixed in 1% glutaraldehyde. (a) Post-fixed in osmium and uranium, embedded in Araldite. (b) No osmium but post-fixed in uranium and embedded in LR White by the cold chemical catalytic method following partial dehydration. (a) was etched in 1% hydrogen peroxide and immersed in sodium metaperiodate to remove osmium. Both sections were incubated with rabbit anti-human IgA and goat anti-rabbit IgG conjugated to 10 nm colloidal gold. The label is loosely dispersed all over (a), but clearly localised to the electron dense deposits in the basement membrane of (b).
(Mag. for both x 20,000).

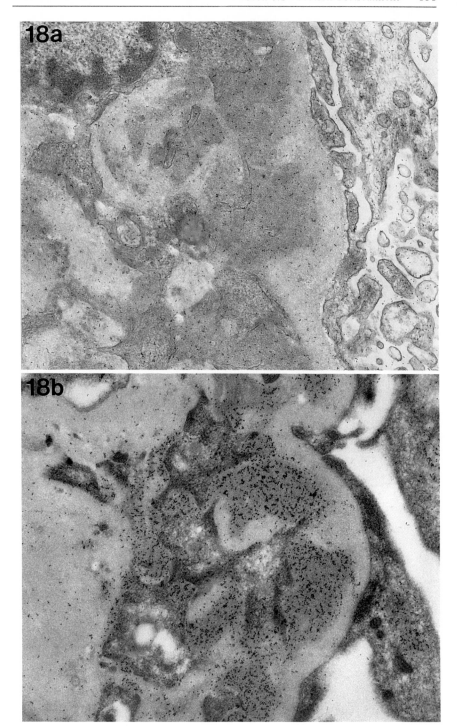

8.1.1.2 Thin Sections

Etching and removal of osmium (Sect. 8.1.5) are again alleged to be helpful in demonstrating some antigens in thin sections of epoxide embedded tissue. The techniques described above are, generally, not suitable for immunoelectron microscopy, although they have been applied to thin sections in a rather convoluted way (Mar and Wight, 1988). Thin epoxide sections are usually etched for 5-10 minutes in 10% hydrogen peroxide (Baskin et al, 1979) which also removes some osmium if it has been used. Care should be exercised when using hydrogen peroxide because it can damage sections (Coggi et al, 1984). Osmium is less detrimentally removed by saturated sodium periodate (Sect. 8.1.5), which, when used for 60 minutes, apparently does not etch but can "unmask" some antigens in thin sections of osmicated, epoxide embedded tissue (Bendayan and Zollinger, 1983). Weak oxidising agents (hydrogen peroxide, periodic acid or sodium periodate) may break hydrophobic ester bonds on the surface of epoxide thin sections to induce a very superficial hydrophilia (Causton, 1984; and personal communication). Both methods very much reduce the contrast of thin sections in the electron microscope.

The etching solution should be washed away thoroughly by passing the grid, either by flotation or immersion, through several droplets of distilled water.

8.1.2 Trypsinisation

Slight improvements in immunoperoxidase staining after trypsinisation of LR White semithin sections have been reported (C. Phillpotts, Department of Science, Bristol Polytechnic, Bristol, UK, personal communication), but trypsin digestion does not seem to promote staining from otherwise negative situations. In fact, immunoglobulins, for example, which cannot be shown in paraffin sections of formalin fixed tissue without trypsinisation, are routinely immunolabelled in untreated LR White sections.

Unosmicated tissue that has been epoxide embedded and etched, however, often reacts to antibodies in a way that is very similar to tissue which has been paraffin wax embedded and rehydrated. For example, the light and heavy chains of immunoglobulins in plasma cells are better demonstrated in etched epoxy resin embedded tissue sections after treatment in a solution of trypsin.

The extent of treatment is a matter for judgement and depends on the level of tissue fixation. Tissue fixed in neutral 10% formalin for long periods of time (>20 hours) will probably need 0.1% trypsin (Sigma) in 0.1% calcium chloride pH 7.8 used at 37°C for 30 min. Tissue heavily cross-linked by glutaraldehyde may need longer. Tissue more lightly fixed in formalin or formalin/glutaraldehyde mixtures may only require 0.05% trypsin for 5-15 minutes. Protease treatment (trypsinisation) is ineffective on epoxide sections when the resin, as in thin sections, is not at least partially removed. Sections should be washed in PBS or TBS to stop trypsinisation (buffer recipies in Sect. 8.2.2).

8.1.3 Inhibition of Endogenous Peroxidase (*Immunoperoxidase only*)

The application of various acids, hydrochloric, picric and periodic, to tissue sections produces a non-specific inhibition of endogenous peroxidase by simply denaturing the protein, but is very damaging, particularly to acrylic thin sections. High concentrations of hydrogen peroxide (3-10%) in water or buffer (PBS) will often only partially inhibit endogenous peroxidatic activity and may also be deleterious to the structure and antigenic sensitivity of tissue. A gentler more specific method employing 0.005% hydrogen peroxide in methanol suppresses peroxidatic activity but can still lead to reduced antigen reactivity (Hittmair and Schmid, 1989). It cannot be used with acrylic sections which are soluble in methanol. There are non-deleterious methods which, when applied to paraffin and frozen sections or to whole cells, have been shown to be very effective although the application of these methods to etched epoxy resin sections or untreated acrylic sections has not been extensively explored. Some initial results have been very promising. They depend on the combination of phenylhydrazine with tiny amounts of hydrogen peroxide (Jasani et al, 1986; Wynford-Thomas et al, 1986). An improved version employs sodium azide in place of phenylhydrazine (Andrew and Jasani, 1987). A third approach uses cyclopropanone hydrate and is even more effective (Schmid et al, 1989). Sections should be washed in PBS or TBS.

8.1.4 Abolition of Aldehyde Groups

Aldehyde groups can be left in tissue from fixation, even following its thorough washing, and may have two major consequences. Primary and secondary reagents, used in cytochemistry and immunocytochemistry, can be attracted to them non-specifically and silver may be deposited by them during photochemical visualisation of colloidal gold and immunoperoxidase/DAB (Sect. 8.2.6). Complex proteins such as collagen, with large numbers of amine or lysine residues, can be particularly troublesome.

Aldehydes can be blocked immediately after fixation and washing of tissue with a solution of ammonium chloride (Sect. 3.2.1.1), but this is often detrimental to both tissue structure and immunoreactivity. In addition, such drastic measures are not always necessary. Gentle on-section blocking is possible with glycine in a buffer (0.1M glycine in PBS, or TBS where tissue has been post-fixed in uranium, see Sect. 8.1.6) applied before immunolabelling but, if the problem is persistent, it may be necessary to treat sections with freshly made 0.1% sodium borohydride for 5-10 minutes at 22°C. Care should be observed because borohydride is very reactive and can damage even epoxy resin sections. Its use almost always reduces the cytochemical and immunocytochemical reactivity of the tissue. Sections should be washed in PBS or TBS. In many cases, though, the addition of bovine serum albumin (BSA) or ovalbumin (OA) to the PBS or TBS buffer may well be adequate to block free aldehyde groups (Sect. 8.1.7).

8.1.5 Osmium Removal

Osmium post-fixation often either completely destroys tissue immunoreactivity or masks it. Photochemical intensification of immunocolloidal gold or DAB is also impractical in the presence of osmium. Some antigenic response may sometimes be restored if osmium is at least partially removed from tissue sections by the use of weak oxidising agents (Baskin et al, 1979; Bendayan and Zollinger, 1983). 10% v/v hydrogen peroxide for 15 to 60 minutes has been advocated for semithin and thin epoxy resin sections (Baskin et al, 1979) although this may result in "etching" of the section surface and damage to the tissue (Coggi et al, 1984). This method is too detrimental for successful use on acrylic sections. Saturated, aqueous sodium metaperiodate for 60 minutes is alledged to remove osmium and unmask antigenic response without etching (Bendayan and Zollinger, 1983), though recovery of immunoreactivity is not optimal (Bendayan, 1984). Periodate also destroys carbohydrate so lectin cytochemistry is impossible following its use. When employed on acrylic resin sections care should be excercised or the periodate will completely remove glycoconjugates weakening thin sections or leaving holes in them. Removal of osmium severely reduces tissue contrast in thin epoxy resin sections and makes them difficult to counterstain (see Sect. 8.2.7). Following osmium removal, sections should be thoroughly washed in several changes of distilled water, then PBS or TBS.

8.1.6 Uranium

Tissues post-fixed in uranyl acetate raise a different problem. Uranyl acetate usefully increases membrane contrast and probably preserves lipid to give improved ultrastructure (Glauert, 1975). It is much less detrimental to antigenic reactivity than osmium (Berryman and Rodewald, 1990; Erickson et al, 1987). However, when pretreating or immunolabelling acrylic resin sections of uranium fixed tissue with reagents containing phosphate buffer, precipitates often give an unsightly background of electron dense spicules and interfere with photochemical visualisation methods. These precipitates can persist even though the tissue, prior to embedding, has been thoroughly washed. Phosphate buffers, therefore, have to be avoided and should be replaced with Tris/HCl buffers which do not cause such problems. Sections of uranium post-fixed tissue should be equilibrated in Tris/BSA, OA or gelatin as appropriate (Sect. 8.2.2 for recipies), and immunolabelled in solutions of primary and secondary immunoreagents diluted in the equilibration solution. 0.02% sodium azide can be added as a bactericide (caution sodium azide is toxic) for Tris buffers are notorious for supporting bacterial growth. They should be kept deep-frozen when not in use and, when at room temperature frequently inspected because they have a limited shelf-life.

8.1.7 Equilibration

It is good practise to equilibrate sections for a few minutes in a buffer containing a non-specific protein such as albumin or gelatin (the choice of which is discussed in Sect. 8.2.2). This solution should be the same as the one chosen for diluting the primary antiserum (Sect. 8.2.2). Equilibration is beneficial even after other on-section pretreatments, and, with untreated sections, serves to hydrate the section (acrylic sections will swell considerably in aqueous solution). The protein inclusion helps to block any non-specific affinities in the tissue and resin, for example residual aldehyde groups and electrostatic charges. Equilibration and all the following steps in immunolabelling, except jet-washing (Sect. 8.2.3), should be conducted in a humidity chamber. Any clear plastic box with a lid in to which wet tissue can be incorporated would be suitable to prevent evaporation from the small drops of immunoreagents (35-50 μl) used on glass slides or with grids.

8.2   Resin Section Immunolabelling

8.2.1 Specific Blocking

Only if high background is a problem, inspite of all efforts (see Sects. 8.1.4/8.1.7), should specific blocking with non-immune sera or purified immunoglobulin fractions be tried. Paradoxically, the incorrect or gratuitous use of these can actually contribute to high background. It can also result in serious reduction or even complete abolition of immunolabelling. Blocking is of two kinds. In inhibitive blocking (pre-blocking), the blocking reagent is applied on-section prior to incubation in primary or secondary reagents. Competitive blocking occurs when the primary and/or secondary reagents are diluted in the blocking reagent. For very heavy blocking, both can be used together. A high concentration of heterologous normal swine (NSS - 1:5 in PBS or TBS) is often used as a blocking agent.

Over use of heavy blocking can result in a serious reduction of specific staining. In any event, low-affinity, non-specific staining caused by the primary reagent can often be diluted out. It is sometimes the secondary detection system that is responsible for high background staining, because it is always used in excess amounts. Background can, in this case, be reduced by further diluting it or, if this threatens the quality of staining, by employing the common manoeuvre of exposing the section to non-immune serum (inhibitive blocking) taken from the species of animal whose immunoglobulins are conjugated to the colloidal gold or peroxidase detection system. The most common of these are normal goat serum (NGS) when goat anti-rabbit or goat anti-mouse gold conjugates are in use, and normal **sheep** serum (NShE) and normal **rabbit** serum (NRS) for peroxidase conjugates. Its effect is limited to the secondary detection system, by applying the blocking agent after the primary reagent incubation. In order to understand the effect of the blocking agent, completely immunolabelled sections should be compared blocked and unblocked. In addition, blocked controls, such as those from which the primary reagent has been omitted, should be compared with unblocked controls.

Non-immunocytochemical secondary detection systems such as Protein A- and avidin/streptavidin (Chap. 7) may need dilutions of unconjugated protein A, avidin or streptavidin to block unspecific staining but these must, of course, be employed before the section is incubated in the primary reagent.

The blocking agent should be washed off in PBS or TBS.

8.2.2 The Primary Reagent

To conserve expensive chemicals, the primary reagent is applied in as small a droplet as is practicable. Depending on size, semithin sections can usually be covered by 50 µl and grids can be floated on or immersed in even smaller amounts provided that the reagent droplets are placed on a hydrophobic surface such as dental wax or 'Parafilm' contained at a high humidity (Sect. 8.1.7). Immunolabelling thin sections by grid flotation has been preferred for epoxy resin sections, alledgedly to reduce background, although quite the reverse is true of acrylic resin sections for which immersion staining is strongly recommended. If attempting the rather impractical double immunolabelling method advocated by Bendayan (1984), in which each side of the grid is labelled with a different primary and secondary reagent, the 'two-faced' method, flotation techniques are at a premium.

When first employing a new primary reagent, such as an antiserum or lectin, a series of dilutions is essential. The diluting agent, (unless the tissue was prefixed in uranyl acetate; see below, this section), is usually 0.01M phosphate buffered saline/bovine serum albumin (PBS/BSA) to which is added sodium azide as a bactericide; (caution should be used because sodium azide is a toxic chemical). PBS/BSA is made as follows:-

0.2M NaH_2PO_4	0.23 g	(or 0.3 g .$2H_2O$)
0.2M $Na_2HPO_4.2H_2O$	1.44 g	(or 2.17 g .$7H_2O$ or 2.9 g .$12H_2O$)
9 g NaCl		
1 L distilled water		

The pH is adjusted to 7.4 after the addition of 0.6 g of BSA and 0.02 g of sodium azide to each 100 ml. All diluents must be either spun at high speed on a bench microcentrifuge (e.g. Eppendorf 320) or microfiltered (e.g. Millipore, 0.22 μm pore size, Millex-GS filter units) in order to remove solid contaminants.

It is important to dilute out the low affinity non-specific elements, particularly from polyclonal antisera, and it is convenient to restrict the time of incubation to 1-2 hours. To establish the staining characteristics of a new primary reagent, doubling dilutions of the neat reagent can be set to cover high and low ranges, for example a range of 1/100, 1/200, 1/400, 1/800 and 1/1600 would not be untypical for a polyclonal antiserum and 1/10, 1/20, 1/40, 1/80, 1/160 for a monoclonal. Results from these series will give enough information to say whether extending the dilution profile would be appropriate. Sometimes heavily diluting the primary reagent and extending the time of incubation to, say, overnight at 4°C will lead to big reductions in non-specific background particularly with epoxide sections (Sternberger, 1979).

Other factors can affect the choice of protein added to the diluent. It would be unwise to include BSA if the antigen from which the primary antiserum is raised is bovine, as for example in a number of antisera raised against bovine insulin. Antibovine antibodies may interact with BSA and cause high background. In this case the basic formula for phosphate buffer and azide remain but the 0.6% BSA should be replaced with 0.5% ovalbumin (OA; Calbiochem: Albumin, Egg (Chicken), 5x Crystaline). OA is not a good choice if the antiserum has been raised against chicken antigen and if this is to be used on bovine tissue it would be advisable to avoid both BSA and OA and use 0.1% purified fish gelatin (Sigma) in phosphate buffer/azide instead.

When working with sections of uranium post-fixed tissue (see above and Sect. 8.1.6) phosphate buffer must be avoided and can be replaced with 20 mM Tris/HCl, pH 7.5. TBS/BSA is made as:-

> 0.24 g of Tris
> 0.9 g sodium chloride
> 100 ml of distilled water
> The pH is adjusted to 7.5 by the dropwise addition of 1N HCl.

To 100 ml of the buffer is added 0.6 g of BSA (or other protein if necessary; see above) and 0.02 g of sodium azide and the pH readjusted. As for PBS/BSA it must be either spun at high speed on a bench microcentrifuge (e.g. Eppendorf 320) or microfiltered (e.g. Millipore, 0.22 μm pore size, Millex-GS filter units) in order to remove solid contaminants.

8.2.3 Washing

Between steps semithin and thin sections will need to be cleared of the preceeding reagent before application of the next. For most semithin resin sections adhered to glass slides, jet-washing with PBS from a plastic squee-gee bottle is all that is necessary. Resin sections of uranyl acetate post-fixed tissue can be jet-washed in TBS. Slides may also be placed in suitable containers of washing solutions such as Coplin jars. Most of the washing solution is wiped away from around sections with an absorbent tissue before the next reagent is applied. Acrylic sections in particular shed solution rapidly so care should be observed not to let the sections dry out at any time during immunolabelling.

Jet-washing is also recommended for epoxy resin thin sections to reduce background. It is, of course, yet another subjective measure since too vigorous a wash will also remove specific staining and may even damage the section.

Jet-washing is not necessary for acrylic resin thin sections. The simple passage of the grid through two or three droplets of washing solution is all that is needed. (In general, grids that are being immunolabelled by flotation are jet-washed, whilst those being treated by immersion are passed through droplets of solution).

8.2.4 The Secondary Detection System

The conservation and application of the secondary detection system to sections is as has been described for the primary reagent.

Colloidal gold and peroxidase conjugates can be diluted in PBS/BSA for LM (except with uranium post-fixed tissue, see above, Sect. 8.2.2), but for EM colloidal gold conjugates should be diluted in 20 mM Tris/BSA (TB/BSA), pH 8.2:-

> 0.24 g of Tris
> 100 ml of distilled water
> The pH is adjusted to 8.2 by the dropwise addition of 1N HCl.

To 100 ml of the buffer is added 1 g of BSA and 0.02 g of sodium azide and the pH is readjusted. As before it must be either spun at high speed on a bench microcentrifuge (e.g. Eppendorf 320) or microfiltered (e.g. Millipore, 0.22 μm pore size, Millex-GS filter units) in order to remove solid contaminants.

Figure 19. 1% glutaraldehyde perfusion-fixed pancreas of rat. (a) Prepared by full dehydration and embedding in LR White following cold chemical catalytic polmerisation (Sect. 3.3.3). (b) Prepared by PLT and embedding in Lowicryl K4M (Sect. 3.5.3). Both are labelled for rat anionic trypsin with goat anti-rabbit IgG conjugated to 10 nm colloidal gold. The localisation and intensity of labelling are the same in both preparations, demonstrating that once high concentrations of fixative are used (> 1%) then irrespective of the processing protocol used, the distribution of label is the same (Sect. 1.3.2.1). Sections counterstained with uranyl acetate and lead acetate. (Mag. for both x 21,500).

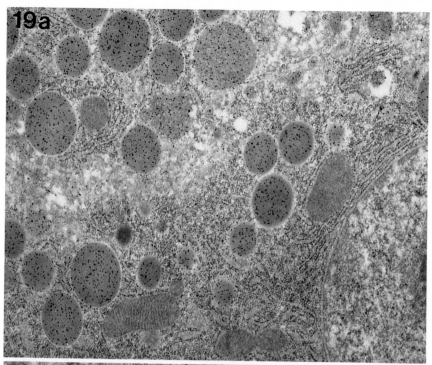

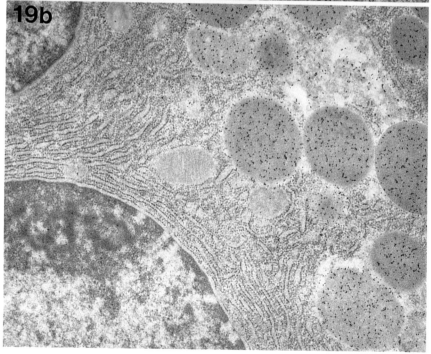

Tris buffer *without* saline is prefered to highly ionic solutions, such as phosphate buffer, as a diluent for colloidal gold when EM immunolabelling, because it does not challenge the colloidal status of the monodispersed gold conjugates. The addition of ions can lead to aggregation of gold particles and increased background. Where the use of a primary antibody against bovine antigen is planned (a good example is ginuea pig antibovine insulin), the inclusion of BSA can sometimes lead to poor labelling. It should be substituted with 0.5% ovalbumin in the TB diluent (see also Sect. 8.2.2).

For indirect methods the concentration of the secondary detection system is theoretically variable but can to a certain extent be rationalised. For epoxide sections it is usually preferable to employ low concentrations of the secondary system (1:100 immunocolloidal gold in TB/BSA, 1:500 immunoperoxidase in PBS/BSA) for long periods of time (4-24 hr) to reduce nonspecific background. This compares with the much higher concentrations and shorter periods of time (1:10 immunocolloidal gold, 1:50 immunoperoxidase for 1-2hr) applicable with acrylic sections, which have little nonspecific attraction for the conjugates.

Fresh sodium azide should not be included as a bactericide in dilutions of peroxidase conjugates because it can inhibit enzymes such as peroxidase.

Those peroxidase methods which require three or four steps (Sect. 7.3) have recommended standardised dilutions and times for all their secondary reagents.

It has been suggested that the addition of 0.1% Tween-20 detergent to the washes and secondary detection systems reduces nonspecific background. This can be useful when immunolabelling epoxide sections but caution should be used when immunolabelling acrylic sections which swell considerably and are easily dislodged or damaged.

8.2.5   DAB (Immunoperoxidase only)

The dry reagent (Sigma) is unstable and should be dissolved in 0.01M PBS as a 0.5% solution. If this solution is suitably aliquoted (e.g. 5 ml aliquots for LM and 0.5 ml for EM) it can be frozen and stored in a deep-freeze (below -20°C) for many months. When required, an aliquot is thawed and diluted ten times (to 0.05%) in strong (0.1M) PBS. DAB can also be dissolved in 20 mM TBS for use with uranyl acetate post-fixed tissue (Sects. 8.1.6/8.2.2). Hydrogen peroxide is added just

---

Figure. 20. Perfusion-fixed pancreas of rat. (a) 0.2% glutaraldehyde perfused for 30 min and prepared by partial dehydration and embedded in Lowicryl K4M with UV-light polymerisation at 0°C (Sect. 3.3.3). (b) 0.1% glutaraldehyde perfused for 15 min and prepared by PLT and embedding in Lowicryl K4M (Sect. 3.5.3). Sections labelled for rat anionic trypsin with goat anti-rabit IgG conjugated to 10 nm colloidal gold. With low fixative concentrations the structure is still adequate, but now low concentration secondary sites in the rough endoplasmic reticulum become labelled (Sect. 1.3.2.1; see also Fig. 19). (Mag. for both x 21,500)

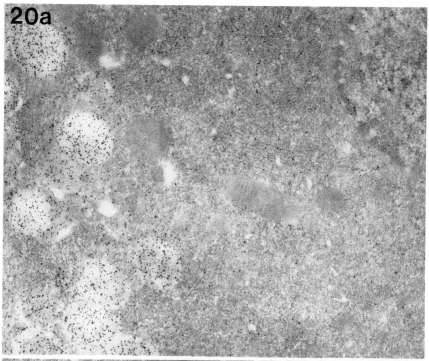

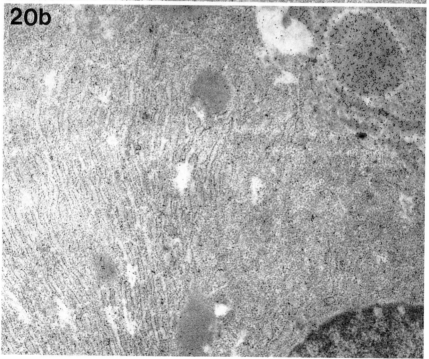

before use (14 ml of 30 vols hydrogen peroxide per 50 ml 0.05% DAB solution; B. Jasani, personal communication). Semi- and thin sections should be left for no longer than 5 minutes in the DAB solution. They are washed thoroughly in at least two 1 minute changes of PBS or TBS followed by at least three 1 minute changes of distilled water to remove unpolymerised DAB.

DAB in thin sections has only a limited electron density. It can be improved with 1% osmium tetroxide (Seligman et al, 1968) but the background tissue in acrylic resin sections may then also take up osmium, masking specific staining. DAB has a great affinity for gold salts, such as sodium gold chloride, which have very high electron density. Used as a 0.1% solution for 1-2 minutes sodium gold chloride dramatically increases the electron density of DAB without affecting the background and is much safer to use than osmium (Newman et al, 1983b).

8.2.6 Photochemical Visualisation of Colloidal Gold and DAB (Silver Intensification)

When used on semithin sections, colloidal gold is frequently completely invisible in normal conditions of viewing in the light microscope. Expensive epi-illumination photomultiplier systems have been used to visualise colloidal gold on sections (De Mey, 1983), but fortunately colloidal gold catalises the deposition of pure silver on its surface when in the presence of a physical developer (one which contains silver ions), thus increasing its bulk and making it visible in the ordinary light microscope (Holgate et al, 1983; Danscher and Rytter Nörgaard, 1983; Springall et al, 1984). Modern, simple and inexpensive methods allow the progress of the visualisation step to be monitored in ordinary daylight conditions under a light microscope (Gallyas et al, 1982). Commercial kits contain complete instructions for use (Intense II, Amersham International; C-Gold, BioClinical Services).

DAB on semithin sections may be scarcely more visible than colloidal gold. It needs to be slightly modified, but then it too is able to photochemically catalyse silver deposition (Newman et al, 1983c) in much the same way as colloidal gold (DAB Intensification Kit, Amersham International).

After the deposition of colloidal gold and prior to or following intensification it is good practice to treat the sections for 5 minutes with 1% neutrally buffered glutaraldehyde to covalently bind the marker in place. If used prior to silver intensification the sections should be thoroughly washed in distilled water or buffer

---

Figure 21. Pancreas of rat prepared by cryomethods. (a) Unfixed tissue rapidly frozen and prepared by cryosubstitution and embedding in Lowicryl K4M (Sects. 4.3.1/4.5.2). (b) 1% glutaraldehyde immersion-fixed tissue prepared by the Tokuyasu method (cryoultramicrotomy; Sect. 4.9). Both are labelled for rat anionic trypsin with goat anti-rabbit IgG conjugated to 10 nm colloidal gold. Only in the cryosubstituted sample are the low concentration secondary sites localised (see also Figs. 19 and 20). Sections counterstained with uranyl and lead acetate. (Mag for both x 21,500).

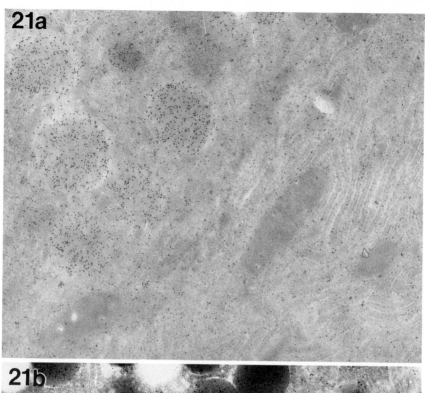

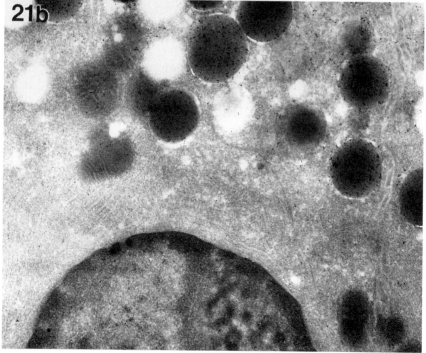

and any residual aldehyde groups eliminated with 0.75 M Tris acetate pH 7.5. (Add 9 g of Tris to 90 ml of water and approximately 3.25 ml of concentrated acetic acid. Adjust to pH 7.5 and make up to 100 ml). The Tris acetate buffer is applied as three one minute changes and the final change is either washed away in distilled water or, if it is compatible with the chosen photochemical visualisation procedure, the excess need only be wiped away from around the sections. The final black stain will now be permanent even after mounting in a synthetic mountant.

The use of very small gold particle sizes (1-3 nm), although possibly having high sensitivity, can result in difficulties of visibility on thin sections in the electron microscope. This is particularly true following counterstaining with heavy metals (Sect. 8.2.7). Once again photointensification can solve the problem by enlarging the gold particles for EM viewing (Stierhof and Schwarz, 1989; Stierhof et al, 1991, 1992). It is important to microfilter all solutions to avoid contamination and a series of development times, of for example 1, 2 and 4 minutes, is advisable so that a time can be selected which does not produce unwanted background silver grains. The photochemical solution is simply washed off in double distilled water to stop the reaction after which the section can be counterstained.

8.2.7 Counterstaining

8.2.7.1 Semithin Sections

Warmed, 1% Toluidine Blue in 1% borax is the standard stain for unetched epoxy resin sections of osmicated tissue. Lower dilutions can be used, without heat, for the more receptive etched epoxy resin, LR White or LR Gold sections of unosmicated tissue. Care should be excercised in the depth of staining or it will mask immunolabelling. Many common laboratory stains can be used on etched epoxy resin and some acrylic sections of unosmicated tissue. Some of the Lowicryl resins (K4M, K11M, but not HM20, HM23) take up basic stains obscuring the embedded tissue (Sect. 7.1), and for these phase-contrast microscopy is probably more productive. 1% Methyl Green is very popular because it is rapid, gives good nuclear definition and contrasts well with the black silver reaction product of photochemical visualisation. Methyl Green is, however, soluble in ethanol and may

---

Figure 22. 1% glutaraldehyde perfusion-fixed pituitary of rat embedded in LR White using the room temperature rapid chemical catalytic method as described for Fig. 1. DNP-labelled rabbit anti-TSH antiserum has been followed by the DHSS peroxidase/DAB method. The DAB has been made more electron dense with 0.1% sodium gold chloride (Sect. 8.2.5). No counterstain. Although at very low magnification the DAB labelled storage granules are clearly visible even when sparsely distributed around the periphery of cells. This low magnification clarity could not be achieved with immunocolloidal gold. See also Fig. 24. (Mag x 7,000).

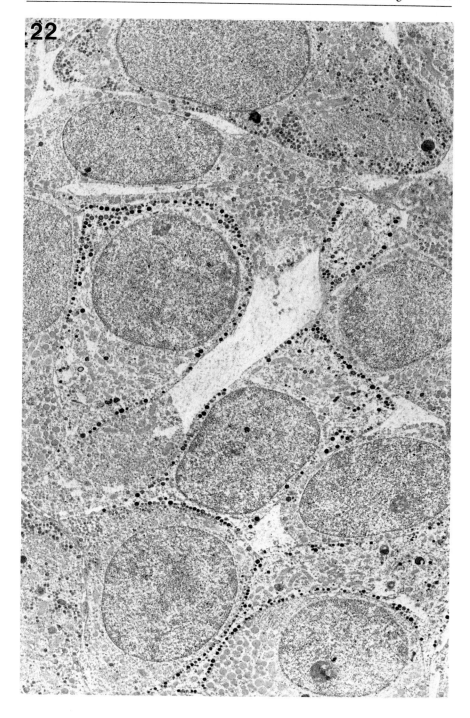

come out of the section if dehydration is necessary (Sect. 8.2.8). Haematoxylin and/or eosin (H&E), also produce an excellent contrast to black silver stains and work well on etched epoxy resin sections and some acrylics. They are diminished in intensity when used on glutaraldehyde fixed tissue and heat may be required (Bowdler et al, 1989). Both Haematoxylin and Eosin stain sections of tissue embedded in LR White more strongly when acetified. Other stains such as Masson's trichrome etc. can also be used.

8.2.7.2 Thin Sections

Heavy metal impregnation to increase contrast is essential in epoxy resin sections. The high electron beam cross-scattering caused by the resin results in poor tissue contrast, especially following the removal of osmium (Sect. 8.1.5). Sections can be single stained in saturated uranyl acetate or nitrate alone, but this is usually quite insufficient to give good contrast. The use of uranium staining solutions must be avoided if the tissue was post-fixed in uranyl acetate before embedding, particularly if the tissue was post-fixed in osmium as well. In these cases only single staining for 2 - 3 minutes in lead citrate (Reynolds, 1963) or lead acetate (Millonig, 1961) is normally necessary. In the absence of uranium post-fixation, single staining in lead is usually insufficient.

Sections can be double stained by floating or immersing them in saturated aqueous uranyl acetate or nitrate for 5 - 60 minutes (depends on the resin) and after washing in several changes of double distilled water, floating or immersing them in either lead citrate or lead acetate for 2 - 5 minutes. Sections of some resins, for example Spurr's resin (Spurr, 1969) are particularly difficult to stain and may need extended times, the application of heat or alcoholic uranyl acetate (4% uranyl acetate in 70% ethanol). If heat is applied to acoholic uranium solutions evaporation of the ethanol can result in on-section precipitates of uranium. Sandwich staining, i.e. lead followed by uranium followed by lead again, can produce very high contrast which could obscure gold markers but may be necessary, for example, following removal of osmium.

Great care must be used when counterstaining acrylic resin sections. They are often extremely receptive and will stain so densely that colloidal gold or DAB markers are completely lost against the background. Dilutions of the solutions used for epoxy resin sections are recommended. Single staining by immersion of sections in

Figure 23. BGPA immersion-fixed surgically removed human pituitary embedded in LR White as described in Fig. 17. DNP-labelled rabbit anti-ACTH has been followed by the DHSS peroxidase/DAB method. The DAB has been made more electron dense with 0.1% sodium gold chloride. No counterstain. The storage granules, endoplasmic reticulum and Golgi, of two kinds of corticotroph are heavily labelled with DAB. (Mag. x 7,000).

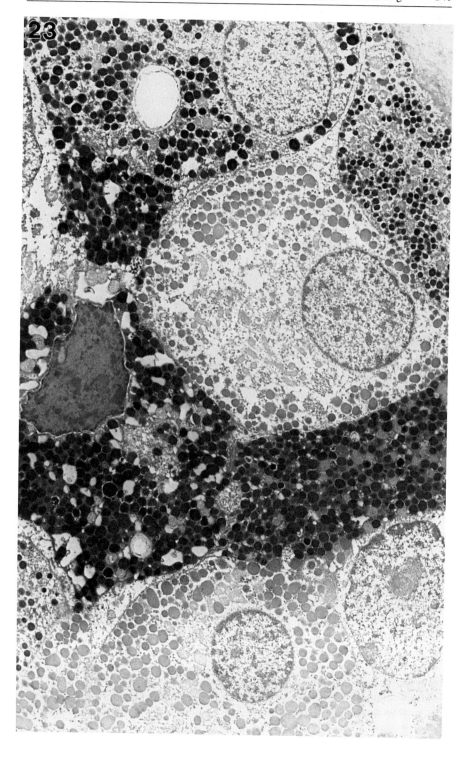

2% aqueous uranyl acetate or nitrate for 3 - 5 minutes followed by three 30 second washes in double distilled water is sometimes all that is needed. Distilled water washes should be rapid because uranium salts are lost from acrylic sections very quickly. However, if the washes are too rapid uranium crystals will precipitate in the tissue on air-drying. Double staining will give more distinct membrane contrast. The uranium salt, used as for single staining, should be washed away quickly in two 45 second changes of distilled water and the section immersed in a dilution of Reynold's (1963) lead citrate or Millonig's (1961) lead acetate. A 1:5 dilution is usually applied for 1 - 2 minutes, but all of the concentrations and times suggested here are empirical and the operator must determine what is most suitable for his tissue and resin. It is very advisable to test a selection of counterstaining procedures on unimmunolabelled thin sections so that a dependable method is available.

Acrylic sections of uranyl acetate post-fixed tissue may still need double staining in uranium and lead. Uranium seems to wash out of tissue, during immunolabelling for example, so that on examination in the EM it has little more contrast than unpost-fixed tissue. Tris buffers are used during the immunolabelling of uranium post-fixed tissue, to avoid the precipitation of uranium crystals (see Sect. 8.2.2), and perhaps this results in its extraction.

8.2.8 Dehydration and Coverslipping of Semithin Sections

Dehydration is only necessary with etched epoxy resin sections and depends on the extent to which the resin was removed. If only the surface of the section was etched, dehydration is not necessary. The section is simply air-dried. If most or all of the resin was removed, however, it will be necessary to dehydrate the section using an ethanolic gradient as one would for deparaffinised sections. Following dehydration, a clearing agent such as xylene should be used before mounting the sections under a coverslip using any proprietary mountant such as DPX or Canada Balsam.

Acrylic sections cannot be dehydrated because they dissolve in high concentrations of organic solvents.

Lightly etched, air-dried epoxy resin sections or any acrylic section may wrinkle if mounted in a xylene containing mountant. Gurr's Neutral Mounting Medium (Merck) has been found to give good visibility, preserve the staining and prevent wrinkling.

8.2.9 Air-Drying of Thin Sections

Epoxy resin sections are robust and can be blotted (the grid is touched delicately by an edge to non-fibrous filter paper [e.g. Whatman No. 50] to allow excess water to be drawn off). When the water, trapped between the jaws of the forceps in which the grid is clamped, has been withdrawn with a piece of non-fibrous filter-paper, the grid can be left on the filter-paper to air-dry.

Acrylic resin sections, particularly if under cross-linked are not so robust and are heavily swollen from immersion in aqueous solution. They should not be blotted or considerable folding, creasing and tearing of sections can occur. The water may be dried out from between the jaws of the clamped forceps as described above, but the grid must then be left in the forceps to air-dry naturally, the section returning gently to its non-swollen form, before it can be released.

8.3   Controls in Immunocytochemistry

8.3.1 Introduction

Following successful immunolabelling, it is important to know if the result can be trusted with respect to its specificity. This is particularly true when examining a new primary antiserum or a new tissue target. To some extent, if some sites have labelled and others along-side have remained negative, the distribution of the immunolabel within the tissue will act as an internal control. Furthermore, the result can be tested by including external controls which have been selected to prove that an immunocytochemical event has taken place and to characterise its specificity. Changes in ambient conditions and reagent quality can alter immunocytochemical results on a routine application basis. As a consequence, controls must accompany all experiments. They cannot be meaningful if done separately at a later date.

8.3.2 Reagent Controls

8.3.2.1 Omitting the Primary Reagent

The most basic control, which should be included in all indirect or bridge methods whatever the chosen marker, is to omit the primary antibody. Negative labelling in this control strongly validates any immunopositivity seen following inclusion of primary antisera. However, areas in which there is evidence of labelling in the control tissue sections may be caused by unspecific attraction for the marker. In this case, dilution of the secondary detection system to extinction should permit removal of any unspecific labelling. Of course there is a limit to dilution below which demonstration of the primary antibody starts to suffer significantly.

Free aldehyde groups, left over from fixation, often because of insufficient washing, are attractive to large proteins such as those used for conjugates. Treatment of sections with glycine buffers and the inclusion of bovine serum albumen (BSA) or gelatin in equilibration solutions (see Sects. 8.1.4/8.1.7) will often neutralise such reactive groups before immunolabelling.

In the situation where dilution is ineffectual in preventing labelling in a negative control, unwanted specific or cross-reactive labelling is often the reason. Immunoglobulin conjugates which bind to specific natural determinants on the primary antiserum, may recognise those determinants in the background tissue. When the conjugate is isologous with the tissue, this can be a very serious problem (Gee et al, 1990). Protein A and G conjugates, with their high affinity for the Fc groups of some immunoglobulins, can bind to them when they occur endogenously. Blocking (see Sect. 8.2.1) can be used to lower background in these cases but sometimes only the use of a hapten method (see Sects. 7.2.3/7.4), making the secondary detection method specific to an artificial non-tissue determinant, will suffice.

Where biotin, a haptenoid, is employed there may still be a problem with endogenous biotin which will reveal itself in control sections from which the primary antiserum has been omitted. Similarly, when employing immunoperoxidase methods, sites of endogenous peroxidase may be revealed and these must be inhibited (Sect. 8.1.3).

Photochemical methods, now commonly used to amplify DAB and visualise colloidal gold, present special problems. Not only should there be a control omitting the primary antiserum but also one omitting both the primary antiserum and the secondary or bridge detection system. This second control reflects the level of reactivity of the tissue with the silver developer (argyrophilia) which can result from different influences. The argyrophilia of well-washed, chemically well-fixed native tissue components is usually relatively low and requires strenuous application of the kind of physical developers (incorporating both reducer and silver salt) used for amplification (Sect. 8.2.6). This provides a comfortable window of 30-45 minutes in which to demonstrate the marker, either colloidal gold or DAB. However, very lightly fixed tissue can produce unexpectedly high background, especially with prolonged silver amplification. A number of amino acid residues, such as lysine, remain reactive with silver developers if they are not thoroughly cross-linked during fixation. More seriously, tissue/aldehyde combinations can be highly argyrophilic and indeed the condensation product of 5-hydroxytriptamine and formaldehyde, a fluorescent iso-carbylamine, will reduce silver salts to silver metal *in situ* without the need for an externally applied reducing agent (argentaffinity). Clearly, taking in to account the attraction that aldehyde groups also have for proteins and some chromogens such as DAB, thorough washing of tissue following fixation, particularly with glutaraldehyde, is very necessary and the treatment of tissue with ammonium chloride (Sects. 3.2.1.1/8.1.4) to abolish aldehydes completely before embedding has been advocated, even though it can be deleterious to ultrastructure. If gentle on-section methods for neutralising aldehyde groups, such

Figure 24. Pitiuitary of rat embedded in LR White as described in Fig. 22. DNP-labelled rabbit anti-ACTH has been followed by the DHSS method as described in Fig. 23. No counterstain. (Mag. x 16,000).

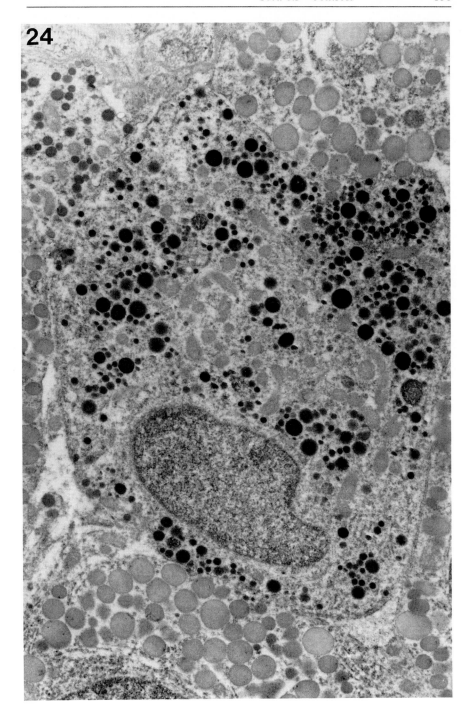

as glycine buffers, BSA and gelatin, fail, the more drastic one of chemically altering them with sodium borohydride (Sect. 8.1.4) will have to be used. Some reduction in the tissue's response to antisera will then almost certainly occur.

8.3.2.2   Dilution   Profiles

A clean control following omission of the primary antiserum, although very desirable, does not entirely validate an immunocytochemical reaction. At high concentrations immunoglobulins will stick to tissue non-specifically because of their natural adhesive qualities. Dilution is, therefore, a very important factor in establishing the affinity of the labelling and a dilution profile of the antiserum using doubling dilutions on serial sections (see Sect. 8.2.2) will give a range over which the labelling persists. It will also indicate where, in the tissue, lie the sites of highest affinity since these will persist longest. Caution, however, should be exercised with bridge systems which will show a bell-shaped curve of immunoreactivity (Bigbee et al, 1977) resulting from exhaustion of the bridge reagent at high primary antiserum concentrations at one end and natural loss of reactivity, which can be quite steep, from over dilution of the primary antiserum at the other. For example, at a tissue level, when using bridge systems, whole cell populations may differ from one another in their affinity for a given antiserum and will be shown by entirely different curves, which possibly may not even overlap,

8.3.2.3   Inappropriate   Primary   Antiserum

In order to show that the immunocytochemical reaction is genuine, a control employing a primary antiserum of the same animal, class and concentration as the primary antiserum under investigation, but which has no expected immunocytochemical affinity with the tissue, should be included. This inappropriate primary antiserum control should also be completely clear of reaction product. "Pre-immune" serum (serum taken from the animal prior to immunisation) has been suggested as an inappropriate primary antiserum control but is actually an omission

---

Figure 25. Immunolabelling of bacteria with 10 nm protein-A gold. (a) *Bacillus subtilis* fixed with 1% glutaraldehyde and preapared by PLT and embedded in Lowicryl HM20 (Sect. 3.5.3). (b) Unfixed *Escherichia coli* cells rapidly frozen and prepared for cryosubstitution in 3% glutaraldehyde/acetone and embedded in Lowicryl K4M at -35°C (Sects. 4.3.1/4.5.2). In (a) the cells are labelled for a cell wall lytic enzyme (N-acetylmuramyl-L-alanine amidase) which is localised to the cell wall, septa and cytoplasm of the bacterium. Counterstained with uranyl acetate. (Mag. x 41,100). In (b) the cells are labelled with a mouse monoclonal specific for ssDNA, which is localised to the periphery of the bacterial nucleoid (ribosome-free areas). Counterstained with uranyl and lead acetate. (Mag. x 38,250).

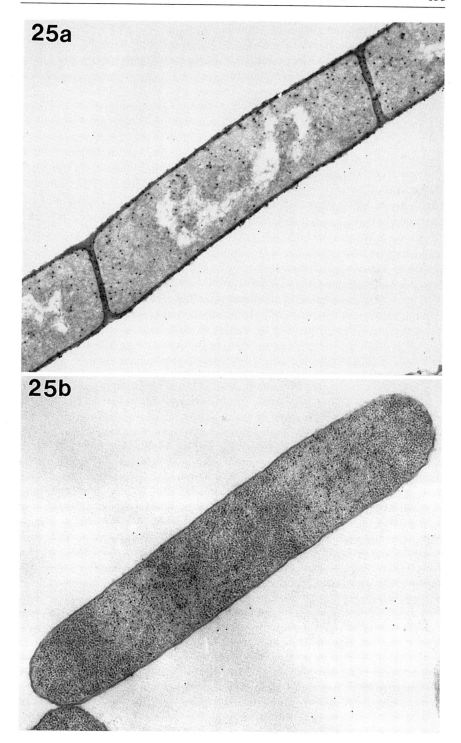

of primary antiserum control. Pre-immune serum may be a useful blocking agent, when used at an appropriate dilution. As with any isologous blocking serum, when used in high concentration, pre-immune serum can compete with the primary antiserum and reduce its effectiveness.

8.3.2.4 Pre-Absorption of the Primary Antiserum

The specificity of some primary antisera can be tested by "pre-absorption" controls. Before use, the antiserum is mixed with a very high concentration of the purified antigen against which it was raised. Since, theoretically, pre-absorption is reversible, especially in liquid phase, it has been suggested that the antigen should be at a concentration which is approximately 100 times that of the antiserum. This rule of thumb is a restraint which can make pre-absorption controls expensive or impracticable and, therefore, they cannot be regarded as routine inclusions. In addition, absorption is not always complete, even at the suggested high concentrations of antigen, and some residual labelling usually remains, especially with small peptide antigens. In many cases, notably with monoclonal antibodies, the antigen against which the antiserum was raised is unknown. This results from the multi-production methods used for raising monoclonal antisera against complex tissue and then screening them for their reactivity with specific tissue elements. Obviously in these situations pre-absorption would be impracticable.

8.3.3 Tissue Controls

Following unsuccessful attempts at immunolabelling questions can arise as to the efficacy of the immunoreagents. Such doubts can be dispelled by the inclusion of a positive control, where possible. This can take the form of a tissue section which is known to produce immunoreactivity with the antiserum and immunomethod in use. It is not entirely necessary to use a section of the same type as that under test. In fact a model such as an immunoblot on nitrocellulose may suffice, as long as it has been shown to produce recognisable immunoreactivity at the same concentration of primary antiserum as the test section. Even if a tissue control for the primary antiserum cannot be constructed, a partial control using a different antiserum but the same secondary detection method will validate some of the reagents involved.

Areas of a tissue left naturally unlabelled by an immunocytochemical reaction coupled with reagent controls will often be enough to characterise a positive immunocytochemical result, but occasionally some room is still left for doubt. For example, if all the tissue immunolabels leaving little of it negative from which to draw comparisons, reagent controls may not be sufficiently credible. In these cases and to further validate reagent controls, truly negative tissue controls can be very useful. The tissue should be chosen for its known lack of antigen specific to the primary antiserum. To allow direct comparisons, it must be processed and resin embedded in exactly the same way as the tissue under investigation.

9 Immunolabelling Protocols for Resin Sections

The protocols for immunolabelling have been set out in three main parts:

    (1) Pretreatment
    (2) Incubation
    (3) Visualisation

In general part (1) is common to all protocols for colloidal gold or immunoperoxidase procedures. Part (3) is common to either all colloidal gold or immunoperoxidase procedures. It is part (2) that varies from procedure to procedure. Therefore the strategy for planning an immunolabelling protocol is to initially select the appropriate labelling technique required (with Incubation [2]), after which parts (1) and (3) are selected.

9.1 Pretreatment Protocols

The pretreatment procedures that are chosen to optimise the section's immunoresponse will depend on the tissue fixation, the embedding resin employed and on the section thickness. The procedures outlined below are common to all immunocytochemistry methods, immunocolloidal gold and immunoperoxidase, although not all will be necessary for any given immunocytochemical or cytochemical reaction. Some are *strongly recommended (in bold italics)* and some can only be chosen after experimentation and, therefore, *may be required (in plain italics)*.

The operational details of these procedures are described in Chapter 8 under Resin Section Pretreatment, Sect. 8.1.

9.1.1 Semithin Sections

All the steps below are completed at 22°C (room temperature) unless otherwise stated, on 350 nm to 2 μm thick sections that are adhered to glass slides. The reagents can be individually jet-washed off with the relevant rinsing solution (PBS, Tris buffer or distilled/deionised water) held in plastic squee-gee bottles, or several slides can be washed together for example in a Coplin jar. Excess washing solution is then wiped away from the vicinity of the section before the next reagent is applied. The sections must never be allowed to dry out. Acrylic sections in particular may shed aqueous solutions very rapidly so care must be taken when wiping slides. An aliquot of the next reagent should be ready to hand so that it can be quickly applied to the section following wiping.

9.1.1.1 Epoxy Resin Section Pretreatment

Etch (15 seconds to 15 minutes)
The resin is either partially or totally removed, with sodium ethoxide or methoxide, revealing the tissue which responds much like that of a rehydrated, deparaffinised section.
Jet-wash with water, then wash with water (3 x 1 minute)
Remove osmium (60 to 180 minutes)
Osmium which can mask or obliterate antigenicity is removed with sodium metaperiodate.
Wash with water (3 x 1 minute)
Trypsinise (5 to 30 minutes)
Immunoglobulins, for example, may be difficult to demonstrate in heavily fixed tissue without trypsinisation.
Wash with PBS or TBS (3 x 1 minute)

9.1.1.2 Epoxy and Acrylic Resin Section Pretreatment

Abolish aldehydes (1 to 10 minutes)
Aldehyde groups left in the tissue from fixation cause high background labelling and interfere with photochemical visualisation. They can be blocked with glycine buffers or chemically removed with sodium borohydride.
Wash with PBS or TBS (3 x 1 minute)
Equilibrate (10 minutes)
Non-specific blocking with large basic protein molecules such as bovine serum albumin (BSA), ovalbumin (OA) or gelatin (G) can reduce non-specific background and all protocols should begin with an equilibration of the section for 5-10 minutes in one of these contained in PBS or TBS.
Block Specifically (10 minutes)
The primary reagent may have an unspecific affinity for some aspect of the tissue. Specific unwanted labelling, which is sometimes a problem with monoclonal antibodies, cannot be blocked.
Wash with PBS or TBS (3 x 1 minute)

9.1.2 Thin Sections

All the steps below are completed at 22°C (room temperature) unless otherwise stated on 50-100 nm thick sections adhered, preferably unsupported, to nickel or gold grids. Sections may be either floated on droplets of reagent (often advocated for epoxy resin sections) or immersed in them. Washing is either by jet for epoxy resin sections or by simple passage through droplets for acrylic resin sections.

9.1.2.1 Epoxy Resin Section Pretreatment

Etch (1 to 15 minutes)
Weak oxidising agents such as hydrogen peroxide may expose antigens on the section surface.
Wash with water (3 x 1 minute) and/or
Remove osmium
Sodium metaperiodate does not etch but removes osmium and 'unmasks' some antigens.
Wash with water (3 x 1 minute)

9.1.2.2 Epoxy and Acrylic Resin Section Pretreatment

Abolish aldehydes (1 to 10 minutes)
Aldehyde groups left in the tissue from fixation cause high background labelling and interfere with photochemical visualisation.
Wash with PBS or TBS (3 x 1 minute)
Equilibrate (10 minutes)
Non-specific blocking with large basic protein molecules such as bovine serum albumin (BSA), ovalbumin (OA) or gelatin (G) can reduce non-specific background and all protocols should include an equilibration of the section in one of these contained in PBS or TBS.
Block Specifically (10 minutes)
The primary reagent may have an unspecific affinity for some aspect of the tissue. Specific unwanted labelling, which is sometimes a problem with monoclonal antibodies, cannot be blocked.
Wash with PBS or TBS (3 x 1 minute)

9.2 Immunocolloidal Gold Labelling Protocols

Of the procedures outlined below some are ***strongly recommended (in bold italics)*** and some can only be chosen after experimentation and therefore, *may be required (in plain italics)*.

Operational details of the procedures listed below are given in Chapter 8 under Resin Section Immunolabelling, Sect. 8.2.

Following appropriate section pretreatment (Sect. 9.1), all sections can be incubated in minimal amounts of the primary reagent (35-50 µl) provided that they are kept at high humidity in a humidity chamber (Sect. 8.1.7).

9.2.1 Visualisation

The visualisation methods for LM and EM viewing of immunogold labelled semithin and thin sections respectively are described in the non-hapten Direct Immunocolloidal Gold method below (Sect. 9.2.2.2). They will not be repeated for the other immunogold procedures.

9.2.2 Direct Immunocolloidal Gold Labelling

9.2.2.1 Incubation

Semithin sections are covered with, thin sections floated on or immersed in:-
1. Primary reagent conjugated to colloidal gold
The conjugate is diluted in Tris (TB) buffer containing BSA, OA or gelatin and the sections incubated for 1 to 24 hours. Short incubations (1-4 hours) are at room temperature, long incubations at 4°C. PBS should be avoided to protect the colloidal status of the gold.
Wash with Tris (TB) buffer (3 x 1 minute)

9.2.2.2 Visualisation

Semithin Sections
1. 1% glutaraldehyde to covalently bind gold to section
Wash with water (3 x 1 minute)
2. 0.75M Tris/acetate to eliminate aldehyde groups (3 x 1 minute)
3. Photochemical reagents to visualise colloidal gold (10 to 20 minutes)
Colloidal gold is usually invisible on semithin sections but can be visualised, with photochemistry, as a black deposit of silver metal (Sect. 8.2.6). The progress of the reaction is monitored in the light microscope and stopped, when judged appropriate, by washing off the reagents. It can be recommenced with fresh chemicals if necessary. (N.B. As an alternative, 1% glutaraldehyde can be used after photochemical silver intensification to stabilise the stain. In this case simply wash the sections with distilled water and apply the photochemical reagents. Tris/acetate is not necessary).

Wash with water (3 x 1 minute)
4. Counterstain (1 to 15 minutes)
Wash with water (3 x 1 minute)
5. Air-dry or dehydrate (10 to 20 minutes)
6. Mount and cover-slip

Thin Sections
Wash with water (3 x 1 minute)
1. Counterstain
Wash with water (3 x 1 minute)
2. Air-dry (10 to 15 minutes)

9.2.3 Indirect Immunocolloidal Gold Labelling (IGS, Protein A/G)

9.2.3.1 Incubation

Semithin sections are covered with, thin sections floated on or immersed in:-
1. Primary reagent (1 to 24 hours)
The diluent may be PBS or TBS containing BSA, OA or gelatin. Dilution is important in avoiding low affinity labelling and may relieve the need for blocking. Prolonged incubation (24 hours at 4°C) of sections in low dilutions of primary reagent can reduce background in epoxy resin sections.
Wash semithin sections with PBS or TBS (3 x 1 minute)
Wash thin sections with primary reagent diluent (2 x 1 minute) and TB/BSA (2 x 1 minute), or if diluent was TBS/BSA
Wash with TBS/BSA (2 x 1 minute) and TB/BSA (2 x 1 minute)
Block Specifically (10 minutes)
Only if the secondary reagent has an unspecific affinity for some aspect of the tissue.
Wash with PBS or TBS (3 x 1 minute) then with TB (2 x 1 minute)
2. Colloidal gold conjugate (1 to 24 hours)
The conjugate should be diluted in TB/BSA. As with the primary reagent, the correct dilution of the colloidal gold conjugate can avoid background and the need for blocking. Using an overnight incubation in a low concentration of colloidal gold at 4°C may reduce non-specific background in epoxy resin sections.
Wash with Tris (TB) buffer (3 x 1 minute)

9.2.4 Hapten-Based Immunocolloidal Gold Labelling (Biotin/Antibiotin or Avidin, DNP AntiDNP)

9.2.4.1 Direct Methods

Incubation
<u>Semithin</u> sections are covered with, <u>thin</u> sections floated on or immersed in:-
1. Antihapten/haptenoid antiserum or avidin conjugated to colloidal gold (1 to 24 hours)
If a hapten (DNP) or haptenoid (biotin, digoxigenin etc.) labelled substance (ligand) has been attached to tissue prior to processing and embedding it may be localised directly by an antihapten/haptenoid antibody or, if the haptenoid was biotin, by avidin conjugated to colloidal gold. Dilution is as for the non-hapten Direct Immunocolloidal Gold method (Sect. 9.2.2).
Wash with Tris (TB) buffer (3 x 1 minute)

9.2.4.2 Indirect Methods

Incubation
<u>Semithin</u> sections are covered with, <u>thin</u> sections floated on or immersed in:-
1. Primary reagent labelled with hapten or haptenoid (1 to 24 hours)
Dilution is as for the non-hapten indirect immunocolloidal gold method (Sect. 9.2.3).
Wash <u>semithin</u> sections with PBS or TBS (3 x 1 minute)
Wash <u>thin</u> sections with primary reagent diluent (2 x 1 minute) and TB/BSA (2 x 1 minute), or if diluent was TBS/BSA
Wash with TBS/BSA (2 x 1 minute) and TB/BSA (2 x 1 minute)
Block Specifically (10 minutes)
Only with biotin may the secondary reagent have an unspecific affinity for some aspect of the tissue.
Wash with PBS or TBS (3 x 1 minute) then with TB (2 x 1 minute)
2. Antihapten/haptenoid or avidin colloidal gold conjugate (1 to 24 hours)
Dilution is as for the non-hapten Indirect Immunocolloidal Gold method (Sect. 9.2.3).
Wash with Tris (TB) buffer (3 x 1 minute)

9.3 Immunoperoxidase Labelling Protocols

Of the procedures outlined below some are ***strongly recommended (in bold italics)*** and some can only be chosen after experimentation and therefore, *may be required (in plain italics)*.

The operational details of the procedures listed below are given in Chapter 8 under Resin Section Immunolabelling, Sect. 8.2.

Following appropriate section pretreatment (Sect. 9.1), all sections can be incubated in minimal amounts of the primary reagent (35-50µl) provided that they are kept at high humidity in a humidity chamber (Sect. 8.1.7).

9.3.1 Resin Section Pretreatment

In addition to the pretreatments already listed above in this chapter under Pretreatment Protocols (Sect. 9.1), one procedure specific to immunoperoxidase labelling remains and is normally completed just before equilibration.

9.3.1.1 Inhibition of Endogenous Peroxidase

Often the method used to fix the tissue or, in the case of epoxy resin, to etch the section will also inhibit endogenous peroxidase. Some tissue structures are, however, particularly prone to endogenous peroxidatic or pseudoperoxidatic expression such as the nuclear membrane, Golgi apparatus, lysosomes, peroxisomes, mitochondria etc. and may need supression (for methods prior to washing see Sects. 8.1.2/8.1.3).
Wash PBS or TBS (3 x 1 minute)
Jet-wash semithin sections, wash thin sections in or on droplets.

9.3.2 Visualisation

The visualisation methods for LM and EM viewing of immunoperoxidase/DAB labelled semithin and thin sections respectively are outlined in the non-hapten direct immunoperoxidase method below (Sect. 9.3.3.2). They will not be repeated for the other immunoperoxidase procedures.

9.3.3 Direct Immunoperoxidase Labelling

9.3.3.1 Incubation

Semithin sections are covered with, thin sections floated on or immersed in:-
1. Primary reagent conjugated to peroxidase (1 to 24 hours)
The conjugate is usually diluted in PBS/BSA but may be diluted in PBS or TBS buffer containing BSA, OA or gelatin. Short incubations (1-4 hours) are at room temperature, long incubations at 4°C.
Wash with PBS or TBS (3 x 1 minute)

9.3.3.2 Visualisation

Semithin Sections
1. DAB solution (3 to 5 minutes)
The buffered solution must be freshly made and activated with hydrogen peroxide.
Wash with water (3 x 1 minute)
2. Photochemical reagents to visualise DAB (30 to 60 minutes)
DAB is frequently very faint in semithin sections but can be visualised, with photochemistry, as a black deposit of silver metal (Sect. 8.2.6). The progress of the reaction is monitored in the light microscope and stopped, when judged appropriate, by washing off the reagents. It can be recommenced if necessary.
Wash with PBS or TBS (2 x 1 minute)
Wash with water (3 x 1 minute)
3. Counterstain (1 to 15 minutes)
Wash with water (3 x 1 minute)
4. Air-dry or dehydrate (10 to 20 minutes)
5. Mount and cover-slip

Thin Sections
1. DAB solution (3 to 5 minutes)
Wash with PBS or TBS (2 x 1 minute)
Wash with water (3 x 1 minute)
2. 0.1% gold chloride (2 minutes)
Increases the electron density of DAB
Wash with water (3 x 1 minute)
2a. Counterstain
Counterstaining can obscure DAB. In acrylic sections contrast is often enough without it.
Wash with water (3 x 1 minute)
3. Air-dry (10 to 15 minutes)

9.3.4 Indirect Immunoperoxidase Labelling

9.3.4.1 Incubation

Semithin sections are covered with, thin sections floated on or immersed in:-
1. Primary reagent (1 to 24 hours)
The primary reagent is usually diluted in PBS/BSA but may be diluted in PBS or TBS buffer containing BSA, OA or gelatin. Dilution is important in avoiding low affinity labelling and may relieve the need for blocking. Prolonged incubation of sections (24 hours at 4°C) in low dilutions of primary reagent can reduce background in epoxy resin sections.

Wash with PBS or TBS (3 x 1 minute)
Block Specifically (10 minutes)
Only it the peroxidase conjugate has an unspecific affinity for some aspect of the
section.
Wash with PBS or TBS (3 x 1 minute)
2. Peroxidase conjugate (1 to 24 hours)
The conjugate is usually diluted in PBS/BSA but may be diluted in PBS or TBS
buffer containing BSA, OA or gelatin. As with the primary reagent, the correct
dilution of the peroxidase conjugate can avoid background and the need for
blocking. Using an overnight incubation in a low concentration of peroxidase
conjugate may reduce non-specific background in epoxy resin sections.
Wash with PBS or TBS (3 x 1 minute)

9.3.5 Hapten-Based Immunoperoxidase Labelling

9.3.5.1 Direct

Incubation
Semithin sections are covered with, thin sections floated on or immersed in:-
**1. Antihapten/haptenoid antiserum or avidin conjugated to peroxidase (1 to 24
hours)**
If a hapten (DNP) or haptenoid (biotin, digoxygenin etc.) labelled substance
(ligand) has been attached to tissue prior to processing and embedding it may be
localised directly by an antihapten/haptenoid antibody or, if the haptenoid was
biotin, by avidin conjugated to peroxidase. Dilution is as for the non-hapten direct
method (Sect. 9.3.3).
Wash with PBS or TBS (3 x 1 minute)

9.3.5.2 Indirect

Incubation
Semithin sections are covered with, thin sections floated on or immersed in:-
1. Primary reagent labelled with hapten or haptenoid (1 to 24 hours)
Dilution is as for the non-hapten indirect method (Sect. 9.3.4).
Wash with PBS or TBS (3 x 1 minute)
Block Specifically (10 minutes)
Only with biotin may the secondary reagent have an unspecific affinity for some
aspect of the tissue.
Wash with PBS or TBS (3 x 1 minute)
2. Antihapten/haptenoid or avidin peroxidase conjugate (1 to 24 hours)
Dilution is as for the non-hapten indirect method (Sect. 9.3.4).

Wash with PBS or TBS (3 x 1 minute)
9.3.5.3 DNP Hapten Sandwich Staining Technique (DHSS)

Incubation

Semithin sections are covered with, thin sections floated on or immersed in:-
1. Primary reagent labelled with DNP (1 to 24 hours)
Dilution is as for the non-hapten indirect method (Sect. 9.3.4).
Wash with PBS or TBS (3 x 1 minute)
2. AntiDNP antiserum (1 to 2 hours)
The antiserum is a monoclonal IgM. It is used at a constant dilution of 1:500 of the ascites fluid.
Wash with PBS or TBS (3 x 1 minute)
3. DNP-peroxidase (1 to 2 hours)
The conjugate is used at a constant dilution of 1:800
Wash with PBS or TBS (3 x 1 minute)

9.3.5.4 4-Layer DNP Hapten Sandwich Staining Technique (4-DHSS)

The DHSS is applicable to unlabelled primary antisera by using a species specific antibody labelled with DNP.

Incubation

Semithin sections are covered with, thin sections floated on or immersed in:-
1. primary reagent (1 to 24 hours)
Dilution is as for the non-hapten indirect method (Sect. 9.3.4).
Wash with PBS or TBS (3 x 1 minute)
2. Species specific antibody labelled with DNP (1 to 24 hours)
If the primary reagent was a rabbit polyclonal, a DNP-labelled antirabbit antiserum is used; if a mouse monoclonal a DNP-labelled antimouse antiserum is used. A 'universal' DNP-labelled antiserum which copes with either is available (BioClinical Services).
Wash with PBS or TBS (3 x 1 minute)
3. AntiDNP antiserum (1 to 2 hours)
As for DHSS (Sect. 9.3.5.3).
Wash with PBS or TBS (3 x 1 minute)
4. DNP-peroxidase (1 to 2 hours)
As for DHSS (Sect. 9.3.5.3).
Wash with PBS or TBS (3 x 1 minute)

9.4   EM Double Immunolabelling Protocols

As with all other immunocytochemical and cytochemical procedures double immunolabelling should be preceded by the appropriate section pretreatment(s) (Sect. 9.1).

Of the procedures outlined below some are ***strongly recommended (in bold italics)*** and some can only be chosen after experimentation and therefore, *may be required (in plain italics)*.

9.4.1 Immunocolloidal Gold Labelling

9.4.1.1 Visualisation

The preparation for EM viewing of immunogold labelled thin sections is described in the non-hapten Direct Immunocolloidal Gold method (Sect. 9.2.2.2).

9.4.1.2 Direct Methods

Incubation
Thin sections floated on or immersed in:-
1. A 'cocktail' of primary reagents (2 or more) each conjugated to a different size of colloidal gold particle (1 to 24 hours)
Dilution is as for the non-hapten direct immunocolloidal gold method (Sect. 9.2.2). Direct methods are notoriously insensitive and different primary reagents and sizes of colloidal gold have different sensitivities. Steric hindrance may further complicate things. It could be very difficult to achieve a balance of primary reagent dilutions such as to show them all.
Wash with Tris (TB) buffer (3 x 1 minute)

9.4.1.3 Indirect Methods

Primary reagents from different sources
Combinations of primary reagents can be used provided they can be localised by procedures that do not cross-react.

Simultaneous Localisation
The method is quicker if the primary reagents are mixed and applied together. The secondary detection reagents are also mixed and applied together. Steric hindrance is reduced by these means but cross-reaction can occur unless the primary reagents and

their detection methods are very different. In addition, it can be difficult to achieve a balance which does not favour one primary reagent reaction against the other.

Incubation

Thin sections are floated on or immersed in:-

1. A mixture of primary reagent A, for example a rabbit polyclonal antiserum and primary reagent B, for example a mouse monoclonal antiserum (1 to 24 hours)

Wash thin sections with primary reagent diluent (2 x 1 minute) and TB/BSA (2 x 1 minute) or if diluent was TBS/BSA

Wash with TBS/BSA (2 x 1 minute) and TB/BSA (2 x 1 minute)

2. A mixture of AntiA conjugated to colloidal gold of x particle size, for example goat antirabbit IgG conjugated to 10 nm colloidal gold, and AntiB conjugated to colloidal gold of y particle size, for example goat antimouse IgG conjugated to 20 nm colloidal gold (1 to 24 hours)

Wash with Tris (TB) buffer (3 x 1 minute)

Sequential Localisation

If the primary reagents are localised sequentially the method can be time consuming. When two (or three) different sizes of colloidal gold are used they may differ in their sensitivity and steric hindrance is a problem. An example is given below.

Incubation

Thin sections are floated on or immersed in:-

1. Primary reagent A, for example a rabbit polyclonal antiserum (1 to 24 hours)

Dilution is as for the non-hapten Indirect Immunocolloidal Gold method (Sect. 9.2.3)

Wash thin sections with primary reagent A diluent (2 x 1 minute) and TB/BSA (2 x 1 minute), or if diluent was TBS/BSA

Wash with TBS/BSA (2 x 1 minute) and TB/BSA (2 x 1 minute)

2. AntiA conjugated to colloidal gold of x particle size, for example goat antirabbit IgG conjugated to 10 nm colloidal gold (1 to 24 hours)

Wash with TB/BSA (2 x 1 minute) and TBS/BSA (2 x 1 minute)

3. Primary reagent B, for example a mouse monoclonal antiserum (1 to 24 hours)

Dilution is as for the non-hapten indirect immunocolloidal gold method above (Sect. 9.2.3).

Wash thin sections with primary reagent B diluent (2 x 1 minute) and TB/BSA (2 x 1 minute) or if diluent was TBS/BSA

Wash with TBS/BSA (2 x 1 minute) and TB/BSA (2 x 1 minute)

4. AntiB conjugated to colloidal gold of y particle size, for example goat antimouse IgG conjugated to 20 nm colloidal gold (1 to 24 hours)

Wash with Tris (TB) buffer (3 x 1 minute)

9.4.1.4 Hapten-Based Methods

Direct

If more than one hapten or haptenoid has been attached to tissue, perhaps through different ligands, they can be localised with a mixture of antihapten/haptenoid antisera conjugated to colloidal gold or, if biotin was involved, avidin conjugated to colloidal gold. The protocol is much the same as that described for single hapten-based Direct Immunocolloidal Gold immunolabelling (Sect. 9.2.4.1).

Indirect

The great strength of hapten-based indirect methods is that they enable primary reagents derived from the same source, e.g. the same animal species, to be localised on the same section. For example, two or more rabbit polyclonals, or mouse monoclonals could be localised using secondary detection systems conjugated to different sizes of colloidal gold. Steric hindrance can be a problem.

Localising an Unlabelled Primary Reagent and a Similar Primary Reagent Labelled with a Hapten (Haptenoid)

This method must be sequential (see Sequential Localisation under non-hapten indirect methods above, Sect. 9.4.1.3) with the unlabelled primary reagent localised first. If the unlabelled primary reagent is localised second, the IGS or Protein A/G methods will also recognise natural determinants on the hapten-labelled primary antiserum.

Incubation

Thin sections are floated on or immersed in:-

1. Unlabelled primary reagent A (1 to 24 hours)
Wash thin sections with primary reagent diluent (2 x 1 minute) and TB/BSA (2 x 1 minute) or if diluent was TBS/BSA
Wash with TBS/BSA (2 x 1 minute) and TB/BSA (2 x 1 minute)
Block Specifically (10 minutes)
Wash with TBS (3 x 1 minute)
2. Colloidal gold conjugate, particle size x, for example 10 nm colloidal gold (1 to 24 hours)
The gold will be conjugated to Protein A or G, or for IGS to an antirabbit or an antimouse antiserum
Wash with Tris (TB) buffer (3 x 1 minute)
3. Block spare valencies on the colloidal gold conjugate with normal rabbit or mouse serum diluted 1:5 in primary reagent diluent (10 minutes)
Wash with Tris (TB) buffer (3 x 1 minute)
4. Hapten-labelled primary reagent B which may be from the same animal species as A (1 to 24 hours)
Wash thin sections with primary reagent diluent (2 x 1 minute) and TB/BSA (2 x 1 minute) or if diluent was TBS/BSA
Wash with TBS/BSA (2 x 1 minute) and TB/BSA (2 x 1 minute)

Block Specifically (10 minutes)
Wash with TBS (3 x 1 minute) then with TB (2 x 1 minute)
5. Antihapten/haptenoid colloidal gold conjugate, or if haptenoid was biotin, avidin colloidal gold conjugate, particle size y, for example 20 nm colloidal gold (1 to 24 hours)
Wash with Tris (TB) buffer (3 x 1 minute)

Localising Similar Primary Reagents Labelled with Different Haptens (Haptenoids).
This method need not be sequential.
Incubation
<u>Thin</u> sections are floated on or immersed in:-
1. A mixture of primary reagent labelled with hapten (haptenoid) A, and primary reagent labelled with hapten (haptenoid) B (1 to 24 hours)
Wash thin sections with primary reagent diluent (2 x 1 minute) and TB/BSA (2 x 1 minute) or if diluent was TBS/BSA
Wash with TBS/BSA (2 x 1 minute) and TB/BSA (2 x 1 minute)
Block Specifically (10 minutes)
Wash with TBS (3 x 1 minute) then with TB (2 x 1 minute)
2. A mixture of AntiA conjugated to colloidal gold of x particle size, for example 10 nm colloidal gold and AntiB conjugated to colloidal gold of y particle size, for example 20 nm colloidal gold. If biotin was used avidin colloidal gold conjugate can also be included (1 to 24 hours)
Wash with Tris (TB) buffer (3 x 1 minute)

9.4.2 Immunocolloidal Gold/Immunoperoxidase DAB Combined

The end-product of the immunocolloidal gold and immunoperoxidase DAB reactions are so different that they make a useful combination for EM double immunolabelling. This method has only been used sequentially to demonstrate an unlabelled primary reagent and a hapten-labelled primary reagent. The unlabelled primary reagent has been localised with immunocolloidal gold first, the hapten-(DNP)-labelled reagent second with peroxidase (DHSS). Theoretically, it could be used to semi-simultaneously localise two primary reagents labelled with different haptens. Following immunolabelling, colloidal gold is located on the surface of sections only whereas DAB is found inside the section. This separation of sites for the two labels means that steric hindrance is not a problem as it is with different sizes of colloidal gold.

9.4.2.1 Incubation and Visualisation

<u>Thin</u> sections are floated on or immersed in:-

1. Unlabelled primary reagent A (1 to 24 hours)
Wash thin sections with primary reagent diluent (2 x 1 minute) and TB/BSA
(2 x 1 minute) or if diluent was TBS/BSA
Wash with TBS/BSA (2 x 1 minute) and TB/BSA (2 x 1 minute)
Block Specifically (10 minutes)
Wash with TBS (3 x 1 minute) then with TB (2 x 1 minute)
2. Colloidal gold conjugate (1 to 24 hours)
The gold will be conjugated to Protein A or G, or for IGS to an antirabbit or an antimouse antiserum
Wash with Tris (TB) buffer (3 x 1 minute)
3. Block spare valencies on the colloidal gold conjugate with normal rabbit or
mouse serum diluted 1:5 in primary reagent diluent (10 minutes)
Wash with Tris (TB) buffer (3 x 1 minute)
4. Hapten-labelled primary reagent B which may be from the same animal
species as A (1 to 24 hours)
Wash with PBS or TBS (3 x 1 minute)
5. AntiDNP antiserum (1 to 2 hours)
The antiserum is a monoclonal IgM. It is used at a constant dilution of 1:500 of the ascites fluid.
Wash with PBS or TBS (3 x 1 minute)
6. DNP-peroxidase (1 to 2 hours)
The conjugate is used at a constant dilution of 1:800
Wash with PBS or TBS (3 x 1 minute)
7. DAB solution (3 to 5 minutes)
Wash with water (3 x 1 minute)
8. 0.1% gold chloride (2 minutes)
Increases the electron density of DAB
Wash with water (3 x 1 minute)
8a. Counterstain
Counterstaining can obscure DAB. In acrylic sections contrast is often enough without it.
Wash with water (3 x 1 minute)
9. Air-dry (10 to 15 minutes)

The method could be made slightly simpler by employing an Indirect, Imunoperoxidase method (Sect. 9.3.5.2) to localise the second DNP-labelled primary reagent in place of the DHSS method (Sect. 9.3.5.3).

9.5 Immunocolloidal Gold Labelling Protocols for Cryoultramicrotomy

Immunolabelling of grids is generally carried out by flotation of the grids on reagent droplets, as the "thawed " cryosections are mounted on grids covered with a carbon-coated plastic film (e.g collodion, formvar). Excess solution is shaken off gently at each step to prevent "carry-over", but any drying out of the grid is avoided. The washing steps are by passage of the grids over droplets of reagent, as jet-washing would invariably detach the "thawed" cryosections.

The first three steps, Section Pretreatment (Sect. 9.5.1), Abolition of Aldehyde Groups (Sect. 9.5.2), and Equilibration (Sect. 9.5.3), are common to all protocols, as is the final step, Visualisation (Sect. 9.5.6).

Of the procedures outlined below some are *strongly recommended (in bold italics)* and some can only be chosen after experimentation and therefore, *may be required (in plain italics)*.

9.5.1 Section Pretreatment

Grids with the "thawed" cryosections, which had been placed section side down on agarose/gelatin plates (Sect. 4.9.3), are gently floated off with 100 μl 0.01M PBS containing 0.01M glycine, pH 7.4

9.5.2 Abolition of Aldehyde Groups

Flotation, section-side face down, on 30-50 μl drops of 0.01M PBS containing 0.01M glycine, pH 7.4 (3 x 1 minute and 1 x 10 minute)

9.5.3 Equilibration

0.01M PBS plus 0.6% BSA, pH 7.4 (PBS/BSA) (1 x 10 minute)

9.5.4 Direct Immunocolloidal Gold Labelling

9.5.4.1 Incubation

Wash with TB/BSA (3 x 1 minute)
1. Primary reagent conjugated to colloidal gold

The conjugate is diluted in Tris (TB) buffer containing BSA, OA or gelatin and the sections incubated for 1 hour. PBS should be avoided to protect the colloidal status of the gold.
Wash with Tris (TB) buffer (4 x 1 minute)

9.5.5 Indirect Immunocolloidal Gold Labelling (IGS, Protein A/G)

9.5.5.1 Incubation

The primary reagent is diluted in PBS/BSA (see Sect. 8.2.2 for exceptions). A dilution profile is necesary to empirically find the optimal dilution of serum to use.
1. Incubate primary reagent (1 x 60 min)
Wash PBS/BSA (3 x 1 minute)
Wash TB/BSA (4 x 1 minutre)
2. Colloidal gold conjugate (1 to 2 hours)
The conjugate should be diluted in TB/BSA. As with the primary reagent, the correct dilution of the colloidal gold conjugate can avoid background.
Wash Tris (TB) buffer (2 x 1 minute)

9.5.6 Visualisation

1. Wash double-distilled water (6 x 1 minute)
2. 4% aqueous uranyl acetate (1 x 10 minutes)
3. Wash in double-distilled water, by quick transfer over four drops
4a Methyl cellulose solution (0.2% w/v methyl cellulose [400 c.p.s], 1.8% w/v polyethylene glycol [M.W. 1540], 0.01% w/v uranyl acetate, dissolved in double-distilled water) (1x10 minutes) OR ALTERNATIVELY
4b. Polyvinyl alcohol (2% PVA w/v [MW 10,000], 0.2% w/v uranyl acetate, dissolved in double-distilled water) (1 x 10 minutes)
The ratio of uranyl acetate to polyvinyl alcohol used above can be increased from 1:10, which produces positive contrast, to 1:1 to 10:1 to give negative contrast effects (Tokuyasu, 1989).
Excess methyl celluose or polyvinyl alcohol solution is gently removed with filter paper (Whatman No. 50). The filter paper can be applied either to the sides of the grid which is held in a pair of forceps, or the grid is picked up in a small loop and the excess drop of solution removed by gently drawing the grid across the filter paper. A very thin film is left covering the sections (which are now just slices of fixed tissue), to protect them from the ill-effects of air-drying.
5. Air-drying of grid

10 Resin Embedding and Immunolabelling

The major practical variables involved in resin embedding and immunolabelling have been dealt with in the preceding chapters, but in the discussions of these, other, more theoretical parameters were touched upon. A deeper understanding of such considerations may be important for "fine tuning" the sensitivity that can be obtained when labelling resin sections and for this reason they are more fully examined here.

10.1   Penetration of Label

The method preferred for immunolabelling may well rest on the important question as to whether or not antibodies penetrate in to sections, and, if they do, how far. Antigen at depth within sections may be demonstrated by some methods (e.g. peroxidase techniques; Newman and Hobot, 1987), giving them a sensitivity advantage over others which are able only to stain the most superficial levels. For example, using colloidal gold markers, it has been shown that no penetration of antibody in to sections occurs, as the colloidal gold marker is unable to enter either "thawed" cryosections (Stierhof et al, 1986; Stierhof and Schwarz, 1989) or resin sections (Newman and Hobot, 1987). The gold particles sit on the section surface.

There is evidence to suggest that this is also true of rehydrated paraffin and etched, semithin epoxy resin sections where, for example, the intensity of labelling is not increased by using thicker sections. This appears to be contradicted by suggestions that small particle conjugates of colloidal gold (1-5 nm) are more sensitive, when used with paraffin sections, than their larger counterparts. However, this could be explained, without the need for penetration, by there being a more complete packing (less steric hindrance) of the smaller gold on antigenic sites, and, since photochemical visualisation is necessary, lower detectable background. In addition, labelling intensity does not increase with greater section thickness when using monodispersed immunoperoxidase/DAB methods, which, because of their powers of penetration, are preferred by immunocytochemists for pre-embedding immunocytochemistry, and is, therefore, probably independant of the immunolabelling technique.

The complete opposite is true when immunolabelling sections of LR White embedded tissue with monodispersed peroxidase/DAB. The thicker the section the heavier the labelling. This represents circumstantial evidence for the penetration of freely diffusable immunoreagents in to LR White resin (Newman, 1987). Further, cross-sections of re-embedded, immunoperoxidase/DAB labelled LR White sections, showed that antigenic sites in depth within sections were DAB labelled (Newman and Hobot, 1987). In these experiments, the concentration of DAB and the short length of time for which it was used (3 minutes) were kept constant so this phenomenon cannot be simply explained by build-up or diffusion of polymerised DAB product.

When using immunocolloidal gold, which is not freely diffusable and so does not easily penetrate, increasing the section thickness does not improve labelling density. However, in Lowicryl K4M embedded models, using tissue for which the number of antigenic sites appearing at the section surface could be calculated theoretically, the actual number of colloidal gold particles marking the sites following immunolabelling often exceeded the theoretical predictions (Griffiths and Hoppeler, 1986; Kellenberger et al, 1987). This could be explained by limited penetration of the section surface by the primary antibody/colloidal gold detection system but has given rise to another possible explanation - the theory of surface relief (Kellenberger et al, 1987; Sect. 10.3).

It may be possible to allow for antibody penetration in to acrylic resin sections by aiming for a low cross-link density in the polymer meshwork of the resin itself.

10.2 Resin Cross-Linking

The cross-linked density of acrylic resins can be varied using the suggestions set out for chemical polymerisation of either LR White or the Lowicryls (Sect. 5.3). Less cross-linked resins have been found to give a higher immuno-response, if used in combination with partial dehydration protocols (Newman, 1987). Sections of underpolymerised acrylic resin tend to swell enormously in aqueous solution, depending on the extent of their hydrophilia, and this may be the reason that they are so freely permeable. Unfortunately, the cross-linked density of any given resin is often difficult to quantify and control. In the case of the LR resins, it may depend on how recently purchased they were since these resins contain chemical catalysts which start the process of polymerisation before arrival in the laboratory. "Uncatalysed LR White" is now available (Sect. 5.5) which might solve this problem.

The polymerisation process is responsible for the final cross-linked density of the resin block. UV-light (or blue light) polymerisation is probably the most difficult of the polymerisation sources to use for controlling the extent of resin cross-linking because it diffuses from the outside of blocks to their centres creating a gradient of cross-linking, which is heaviest, at first, at the periphery. This is evened out by completing the cross-linking process. For the Lowicryls, though, varying the amounts of crosslinker to monomer added (Varying the Hardness of the Resins in Sect. 2.2.3.3) could result in blocks after UV-light polymerisation having different cross-linked densities. Heat is more controllable, but probably polymerisation by the use of chemical catalysts and accelerators, whose concentrations can be varied to suit different requirements, is the most versatile and controllable method (Sects. 1.3.4/5.3).

Of course, eventually, resin underpolymerisation will result in the deterioration of block sectioning properties and loss of electron beam stability. The user should never be afraid to experiment wih cross-linked density, however, because the polymerisation process can always be continued with UV-light or heat. In fact, assuming reasonable sectioning properties, the final polymerisation source can be the electron beam itself provided that the section is given time to equilibrate before being subjected to high electron bombardment (see Chap. 6).

Whilst for immunoperoxidase/DAB techniques, the reduced cross-linked meshwork of the polymerised resin increases the sensitvity of labelling obtained by probably allowing greater penetration of antibodies in to the resin sections, the increased labelling seen with colloidal gold markers may be due to a combination of factors. Greater penetration of antibodies near to the resin surface, but still allowing for their attachment by colloidal gold markers, is one possibility. Another is the increased surface relief on the top of sections that may be caused upon sectioning less cross-linked resins. This could potentially have the affect of exposing more antigenic sites for antibody recognition.

10.3   Surface Relief of Sections

The role of some kind of relief or micro-chatter in allowing antibody recognition of exposed antigenic sites on the resin section surface is thought to play an important role in immunolabelling (Kellenberger et al, 1987). The section relief on acrylic resin sections can be as much as 2-6 nm, whilst epoxy resins have a surface relief three times less. This increased surface area could account for the improvements in immunolabelling observed on acrylic sections (Kellenberger et al, 1987). A surface relief can be artificially created on section surfaces of epoxy resins by etching procedures, and immunolabelling can then be on a par with that obtained for acrylic resins (Roth et al, 1981; Bendayan, 1984). Only the background is generally higher. However, in the absence of etching or any other pretreatment of epoxy resin sections, aldehyde fixed tissue embedded in Epon has yielded good labelling over the smooth section surface (Bendayan et al, 1980). It has been suggested that the interaction of the level of aldehyde fixation with tissue can affect the surface relief by altering the preservation of hydration shells (Kellenberger and Hayat, 1991). Results from the use of low levels of glutaraldehyde for fixation and either full or partial dehydration (Sect. 1.3.2) have shown that this is unlikely to be the case (Hobot and Newman, 1991). Surface relief is one of a variety of variables that can affect the final immunolabelling result, but other parameters are initially probably of greater importance (Hobot and Newman, 1991; Newman and Hobot, 1987, 1989).

10.4   Quantitation

The possibility of directly relating the number of gold particles seen on section surfaces following immunolabelling, to the concentration of antigen present in the tissue or subcellular compartments is a very attractive one. It would give morphologists, who are frequently criticised for not being able to put figures on their findings, the levels of credence claimed by biochemists for their results. Several attempts have been made to exploit this possibility (Griffiths and Hoppeler, 1986; Kellenberger et al, 1987; Lucocq, 1992; Posthuma et al, 1987), but variation in the complex interplay between the numerous parameters involved in tissue preparation, resin embedding and immunolabelling, which have been the subject of this book, can lead to very different results in the final immunolabelling counts. Artificial standards, with known amounts of a substance, are difficult to prepare and usually do not relate in their immunolabelling to the tissue with which they are to be compared. All too frequently, the method chosen is that which gets closest to biochemical results which have already been established. Without prior knowledge, immunocytochemistry alone cannot be relied upon to give accurate measurements of amounts of substances within cells. The methods of fixation, dehydration, embedding and polymerisation, and the gold labelling technique itself, cannot be standardised for all tissue, for all problems. Even the quality of primary reagents and secondary detection systems such as protein A-gold (Geoghagen, 1991) or immunoglobulin conjugated gold probes (Geoghagen, 1988; Karmarcy and Sealock, 1991), available from commercial suppliers or produced in the laboratory, vary in their sensitivities, purities and, most importantly, their stabilities, giving day to day variation and making experimental continuity very difficult. Until these problems are solved not even approximate quantitation can be attempted, although, given the correct conditions, comparative immunocytochemistry, using gold label counts, remains a possibility.

10.5   Conclusion

The majority of techniques that can be used for preparing a biological specimen for immunomicroscopical analysis have been covered. The requirement placed on the microscopist to make a deliberate choice of protocols, appropriate to his needs, from a spectrum of resin embedding and immunolabelling methods of increasing complexity has been the guiding principle. This choice is based not on trial and error, but on a strategy, where the effects of differing parameters, interdependant on each other in their actions, finally affect not only the biological tissues' structure and immunoreactivity, but also the interpretation that arises from studying the results. Eventually it is to be hoped that this strategy will lead to a sounder scientific understanding than that provided by empiricism alone. Needless to say, still more work has to be carried out to provide more insights in to the ways that immunonmicroscopy can be improved and widened in its scope of techniques, such that microscopy can continue to play an effective role in biological research.

Appendix I

Examples of Typical Resin Embedding Regimes for Immunocytochemistry

1. Solid Tissue, Pellets and Agar Blocks - Immersion Fixation

Tissue   1-2 mm^3 tissue blocks. **Sect. 3.1**
Comments   *If cells in agar blocks are very lightly fixed, use Regime 2 or 3.*

Fixation   Immersion, 1% glutaraldehyde for 3 hr. **Sect. 3.3.1.1**
Comments   *Fixation by diffusion therefore high concentration, long time.*

Washing   Buffer overnight x 2, or buffer 4 x 1 hr. **Sect. 3.3.1.1**
Comments   *Prolonged washing to remove aldehyde, fixation reversal unlikely.*

Option   Ammonium chloride to remove aldehyde. **Sect. 3.2.1.1**
Comments   *May damage ultrastructure.*

Wash   Distilled water 2 x 15 min
Post-Fixation   2% uranyl acetate. **Sect. 3.3.1.1**
Comments   *Improves membrane definition, may slightly reduce antigenic response.*

Wash   Distilled water 2 x 15 min

Dehydration   Ethanol, partial, room temp. **Sect. 3.3.2.1**
Comments   *Heavily fixed LT/PLT pointless. Partial dehydration reduces extraction.*

Infiltrate   70% ethanol/resin mixture; resin 4 x 20 min or overnight.
                    Sect. 3.3.2.2
Comments   *Resin x 4, less extractive but large blocks may need overnight*

Resin	LR White, LR Gold, Lowicryl K4M. **Sect. 3.3.2.2**
Comments	*LR White easiest to use.*

Polymerise	Heat or chemical catalytic (room temp. or 0°C). **Sects. 3.3.2.3/3.3.3.3**
Comments	*Heat for large blocks, chemical catalytic at 0°C least extractive.*

Further Polymerise	50°C for 1-2 hr. **Sects. 5.3.1.1/5.3.1.2**
Comments	*Blocks may need more cross-linking to be beam-stable*

Semithin Sections	350 nm on chrome-gel glass-slides. **Sect. 6.1.2.1**
Comments	*Economic, flatten and stick better.*

Counter-stain	LR White/LR Gold Tol. Blue, Meth. Green, H&E; K4M phase contrast
Comments	*K4M stains unspecifically with basic dyes.*

Thin sections	80-100 nm nickel or gold grids 300 mesh hex. **Sect. 6.1.2.1**
Comments	*Unsupported on smooth side of grid. Any counterstain but not too heavy.*

Immuno-label	LR White/LR Gold immunoperoxidase or immunogold; K4M immunogold. **Chap. 9**
Comments	*May need silver intensification. K4M stains unspecifically with DAB*

2. Solid Tissue - Light Perfusion Fixation

Tissue	0.5-1 mm^3 tissue blocks. **Sect. 3.3.1.2**
Comments	*Can be any solid tissue from any source*

Fixation	Perfusion, 0.5% glutaraldehyde, for 30 min. **Sect. 3.3.1.2**
Comments	*Fixation by perfusion therefore low concentration, short time.*

Washing	Buffer 4 x 15 min. **Sect. 3.3.1.2**
Comments	*Prolonged washing to remove aldehyde unnecessary.*

Option	Ammonium chloride to remove aldehyde unnecessary. **Sect. 3.2.1.1**
Comments	*May damage ultrastructure.*

Wash	Distilled water 2 x 15 min
Post-Fixation	Optional, 2% uranyl acetate. **Sect. 3.3.1.2**
Comments	*Improves membrane definition, may slightly reduce antigenic response.*
Wash	Distilled water 2 x 15 min
Dehydration	Ethanol, partial, or either LT or PLT. **Sects. 3.3.3.1 (3.4/3.5)**
Comments	*Lightly fixed so LT or PLT can be used - see Regime 3. Partial dehydration also reduces extraction.*
Infiltrate	Ethanol/resin mixture; resin 4 x 20 min or either LT /PLT. **Sects. 3.3.3.2 (3.4/3.5)**
Comments	*70% ethanol/resin mixture essential to prevent osmotic shock*
Resin	LR White, LR Gold, Lowicryls. **Sect. 3.3.3.2**
Comments	*LR white easiest to use for partial dehydration, Lowicryls for LT/PLT*
Polymerise	Chemical catalytic (0°C to -35°C) or UV-light. **Sect. 3.3.3.3**
Comments	*PLT at -35 °C probably least extractive - see Regime 3.*
Further Polymerise	50°C for 1-2 hr, UV-light 24-48 hr room temp. **Sects. 5.3.1.1/5.3.1.2**
Comments	*Blocks may need more cross-linking to be beam-stable*
Semithin Sections	350 nm on chrome-gel glass-slide. **Sect. 6.1.2.1**
Comments	*Economic, sections flatten and stick better.*
Counter-stain	LR White/LR Gold/HM20 Tol. Blue, Meth. Green, H&E; K4M phase contrast
Comments	*K4M stains unspecifically with basic dyes.*
Thin sections	80-100 nm nickel or gold grids 300 mesh hex. **Sect. 6.1.2.1**
Comments	*Unsupported on smooth side of grid. Any counterstain but not too heavy.*

Immuno-
label   LR White/LR Gold/HM20 immunoperoxidase or immunogold;
              K4M immunogold. **Chap. 9**

Comments   *May need silver intensification. K4M stains unspecifically with*
              DAB.

3. Solid Tissue - Very Light Perfusion Fixation

Tissue   <0.5 mm^3 tissue blocks. **Sect. 3.5.2**
Comments   *Can be any solid tissue from any source*

Fixation   Perfusion, 0.2% glutaraldehyde, for 30 min. **Sect. 3.5.2**
Comments   *Fixation by perfusion therefore low concentration, short time.*

Washing   Buffer 4 x 15 min. **Sect. 3.5.2**
Comments   *Prolonged washing to remove aldehyde unnecessary.*

Option   Ammonium chloride to remove aldehyde <u>unnecessary</u>. **Sect. 3.2.1.1**
Comments   *Will damage ultrastructure.*

Wash   Distilled water 2 x 15 min

Post-Fixation Optional, 2% uranyl acetate. **Sect. 3.5.2**
Comments   *Improves membrane definition, may slightly reduce antigenic*
              response.

Wash   Distilled water 2 x 15 min

Dehydration Ethanol/PLT. **Sect. 3.5.3.1**
Comments   *Very lightly fixed so PLT.*

Infiltrate   Ethanol/resin mixture; resin as per PLT. **Sect. 3.5.3.2**
Comments   *Final temp. depends upon resin chosen, see below.*

Resin   Lowicryls K4M (-35°C) or HM20 (-50°C). **Sects. 3.5.3/3.5.4**
Comments   *Lowicryls for PLT*

Polymerise   Chemical catalytic (-35°C) or UV-light. **Sect. 3.5.3.3**
Comments   *UV-light usually.*

Further
Polymerise   UV-light 24-48 hr room temp. **Sect. 3.5.3.3**
Comments   *Blocks may need more cross-linking to be beam-stable*

Semithin
Sections   350 nm on chrome-gel glass-slide. **Sect. 6.1.2.2**
Comments   *Economic, sections flatten and stick better.*

Counter-
stain   HM20 Tol. Blue, Meth. Green, H&E; K4M phase contrast
Comments   *K4M stains unspecifically with basic dyes.*

Thin sections   80-100 nm nickel or gold grids 300 mesh hex. **Sect. 6.1.2.2**
Comments   *Unsupported on smooth side of grid. Any counterstain but not too heavy.*

Immuno-
label   HM20 immunoperoxidase or immunogold; K4M immunogold only. **Chap. 9**
Comments   *May need silver intensification. K4M stains unspecifically with DAB.*

4. Cryosubstitution

Tissue   <0.1 mm^3 tissue blocks. **Sect. 4.1**
Comments   *Can be any tissue from any source*

Fixation   Cryoimmobilisation. **Sect. 4.2**
Comments   *Rapid freezing technology.*

Substitution   Acetone -80°C, 3-4 days. **Sect. 4.5.2.1**
Comments   *No fixatives necessary.*

Infiltrate   Acetone/resin mixture; resin at -35°C to -80°C.
                  Sects. 4.5.2.3/4.5.3.3
Comments   *Remove all traces of acetone.*

Resin   Lowicryls. **Sects. 4.5.2.3/4.5.3.3**
Comments   *K4M -35°C, HM20 -50°C, K11M -60°C, HM23 -80°C*

Polymerise   Chemical catalytic (-35°C) or UV-light. **Sects. 4.5.2.4/4.5.3.4**
Comments   *UV-light usually.*

Further
Polymerise   UV-light 24-48 hr room temp. **Sects. 4.5.2.4/4.5.3.4**
Comments   *Blocks may need more cross-linking to be beam-stable*

Semithin
Sections   350 nm on chrome-gel glass-slide. **Sect. 6.1.2.2**
Comments   *Economic, sections flatten and stick better.*

Counter-
stain   HM20 & HM23 Tol. Blue, Meth. Green, H&E;
                      K4M & K11M phase contrast
Comments   *K4M & K11M stain unspecifically with basic dyes.*

Thin sections   80-100 nm nickel or gold grids 300 mesh hex. **Sect. 6.1.2.2**
Comments   *Unsupported on smooth side of grid. Any counterstain but not too heavy.*

Immuno-
label   HM20 & HM23 immunoperoxidase or immunogold;
                      K4M & K11M immunogold only.
Comments   *May need silver intensification. K4M & K11M stain unspecifically with DAB.*

5. Project Planner

Tissue
Comments

Fixation
Comments

Washing
Comments

Option
Comments

Wash

Post-Fixation
Comments

Wash

Dehydration
Comments

Infiltrate
Comments

Resin
Comments

Polymerise
Comments

**Further
Polymerise**
Comments

**Semithin
Sections**
Comments

**Counter-
stain**
Comments

Thin sections
Comments

**Immuno-
label**
Comments

Appendix II

List of Suppliers

Although every effort has been made to present an up to date list of addresses, the reader's attention is drawn to the fact that company "takeovers" have been very common in the last few years. Fortunately, existing companies have usually retained the original name of the products, such that most items can be traced. Company addresses (mainly for those appearing in the text) are given for the UK or Europe, USA and Japan, but most have subsidiaries/agents world-wide.

EM general:

Agar Scientific Ltd.,
66a, Cambridge Road,
Stanstead,
Essex CM24 8DA,
Great Britain.

Amersham International plc,
Lincoln Place,
Green End,
Aylesbury,
Buckinghamshire HP20 2TP,
Great Britain.

Amersham Corporation,
2636 South Clearbrook Drive,
Arlington Heights,
IL 60005,
USA.

Amersham Japan Ltd.,
Tokyo Toyama Kaikan,
1-3 Hakusan-5-Chome,
Bunkyo-ku,
Tokyo 112,
Japan.

BioClinical Services Ltd.,
Units 1-3 Willowbrook Laboratories,
Crickhowell Road,
St. Mellons,
Cardiff CF3 0EF,
Great Britain.

Dako Ltd.,
16, Manor Courtyard,
Hughenden Avenue,
High Wycombe,
Buckinghamshire HP13 5RE,
Great Britain.

Dako Co.,
6392 Via Real,
Carpinteria,
CA 93013,
USA.

Dako Japan Co. Ltd.,
Hiraoka Building,
Nishinotouin-higashiiru,
Shijo-dori, Shimogyo-ku,
Kyoto 600,
Japan.

Electron Microscopy Labs. Ltd.,
848A, Oxford Road,
Reading,
Berks. RG3 1EL,
Great Britain.

Electron Microscopy Sciences,
321, Morris Road,
Box 251,
Fort Washington,
PA 19034,
USA.

Ernest F. Fulham Inc.,
900, Albany Shaker Road,
Latham,
NY 12110,
USA.

E-Y Laboratories Inc.,
127, North Amphlett Blvd.,
San Mateo,
CA 94401,
USA.

Fisons Scientific Equipment,
Bishop Meadow Road,
Loughborough,
Leicestershire LE11 0RG,
Great Britain.

Fluka Chemicals Ltd.,
Peakdale Road,
Glossop,
Derbyshire SK13 9XE,
Great Britain.

ICN Biomedicals Ltd.,
(for ICN Flow products),
Eagle House,
Peregrine Business Park,
Gomm Road,
High Wycombe,
Bucks. HP13 7DG,
Great Britain.

ICN Biomedicals Inc.,
3300 Hyland Avenue,
Costa Mesa,
CA 92626,
USA.

J. Bibby Scientific Products Ltd.,
Stone,
Staffordshire ST15 0SA,
Great Britain.

Ladd Research Industries Inc.,
P.O. Box 1005,
Burlington,
Vermont 05402,
USA.

London Resin Co. Ltd.,
P.O. Box 34,
Basingstoke,
Hampshire RG21 2NW,
Great Britain.

Merck Ltd.,
Customer Service Centre,
Hunter Boulevard,
Magna Park,
Lutterworth LE17 4XN,
Great Britain.

Nisshin EM Co. Ltd.,
23-9, Araki-cho,
Shinjuku-Ku,
Tokyo,
Japan.

Pelco International,
4595, Mountain Lakes Blvd.,
Redding,
CA 96003,
USA.

Pharmacia Ltd.,
Davy Avenue,
Knowhill,
Milton Keynes MK5 8PH,
Great Britain.
(Pharmacia products cf. Leica).

Polysciences Inc.,
400, Valley Road,
Warrington,
PA 18976-2590
USA.

Polysciences Ltd.,
Handelsstr. 3,
D-6904 Eppelheim,
Germany.

Safety 4 A/S,
Lyngby,
Denmark.

Sigma Chemical Co. Ltd.,
Fancy Road,
Poole,
Dorset BH17 7NH,
Great Britain.

Sigma Chemical Co.,
P.O. Box 14508,
St. Louis,
Missouri 63178,
USA.

Taab Laboratories Equipment Ltd.,
3, Minerva House,
Calleva Industrial Park,
Aldermaston,
Reading,
Berkshire RG7 4QW,
Great Britain.

Taab Agent in Japan:
Chiyoda Junyaku Inc.,
Tokuzumi Building 4F,
No. 10-2 Chome,
Jinbocho,
Kanda,
Chiyoda Ku,
Tokyo,
Japan.

Vector Laboratories,
16, Wulfric Square,
Bretton,
Peterborough PE3 8RF,
Great Britain.

Vector Laboratories Inc.,
30 Ingold Road,
Burlinghame,
CA 94010,
USA.

EM apparatus:

Bal-Tec AG,
Postfach 75,
FL-9496 Balzers,
Liechtenstein.
(for Balzers products)

Balzers High Vacuum Ltd.,
Bradbourne Drive,
Tilbrook,
Milton Keynes MK7 8AZ,
Great Britain.

Bal-Tec Products Inc.,
984, Southford Road,
P.O. Box 1221,
Middlebury,
CT 06762,
USA.

Hakuto Co. Ltd.,
Balzers Division,
P.O. Box 25,
Tokyo Central 100-91,
Japan.

Leica UK Ltd.,
Davy Avenue,
Knowlhill,
Milton Keynes MK5 8LB,
Great Britain.

Leica Inc.,
111 Deer Lake Road,
Deerfield,
IL 60015,
USA.

Leica KK,
Akesaka,
Twin-Tower Building,
12th. Floor,
17-22 2-Chome Akesaka,
Minato-Ku,
Tokyo 107,
Japan.

RMC (Europe) Ltd.,
P.O. Box 408,
Oxford,
Oxfordshire OX2 9AR,
Great Britain.

RMC Inc.,
4400 South Santa Rita Avenue,
Tucson,
Arizona 85714,
USA.

RMC Agent in Japan:
Nippon Automatic Control Co. (NACC),
Sales Office,
No. 5 Kanesaka Building,
10-9 Shimbashi 3-Chome,
Minaio-ku,
Tokyo,
Japan.

Electron Microscopes

Carl Zeiss,
Produktbereich
Elektronenmikroskopie,
D-7082 Oberkochen,
Germany.

Hitachi Scientific Instruments,
Nissei Sangyo Co. Ltd.,
Intelligent Plaza Building,
27-26 Sakamachi,
Shinjuku-ku,
Tokyo 160,
Japan.

Jeol Ltd.,
1418 Nakagami Akishima,
Tokyo 196,
Japan.

Philips Nederland B.V.,
Afd. Analysetechnieken, VB3,
Postbus 90050,
5600 PB Eindhoven,
The Netherlands.

Flow Cytometry:

Beckton Dickinson,
Immunocytometry Systems Europe,
Denderstraat 24,
B-9440 Erembodegem,
Belgium.

Beckton Dickinson,
Immunocytometry Systems,
P.O. Box 7375,
Mountain View,
CA 94039,
USA.

Nippon Beckton Dickinson Co. Ltd.,
Shimato Building,
5-34 Akasaka 8-chome,
Minato-Ku,
Tokyo 107,
Japan.

Coulter Electronics Ltd.,
Northwell Drive,
Luton,
Beds. LU3 3RH,
Great Britain.

Coulter Co.,
Coulter Immunology Division,
440 West 20th. Street,
Hialeah,
FL 33010,
USA.

Japan Scientific Instru. Co. Ltd.,
6-7-5, Higashikasai,
Edogawa-ku,
Tokyo 134,
Japan.
(Coulter Co.).

Ortho Diagnostic Systems Ltd.,
Enterprise House,
Station Road,
Loudwater,
High Wycombe,
Bucks. HP10 9UF,
Great Britain.

Ortho Diagnostic Systems Ltd.,
1001 US Route 202 North,
Raritan,
NJ 08869-0606,
USA.

Ortho Diagnostic Systems Ltd.,
No. 3-2 Toyo 6-Chome,
Koto-ku,
Tokyo 135,
Japan.

References

Acetarin J-D, Carlemalm E (1982) Chemical polymerisation methods for methacrylate. Appendix in Carlemalm et al (1982a), pp 140-141.

Acetarin J-D, Carlemalm E (1985) Lowicryl HM23 and K11M: two new embedding resins for very low temperature embedding. In: Carlemalm E, Villiger W, Kellenberger E (eds) Lowicryl Letters, No. 33. Chemische Werke Lowi GmbH, P.O. Box 1660, D-8264 Waldkraiburg, Germany, pp 2-4.

Acetarin J-D, Carlemalm E, Villiger W (1986) Developments of new Lowicryl resins for embedding biological specimens at even lower temperatures. J Microsc (Oxford) 143:81.

Altman LG, Schneider BG, Papermaster DS (1984) Rapid embedding of tissues in Lowicryl K4M for immunoelectron microscopy. J Histochem Cytochem 32:1217.

Amako K, Meno Y, Takade A (1988) Fine structure of the capsules of Klebsiella pneumoniae and Escherichia coli K1. J Bacteriol 170:4960.

Amako K, Murata K, Umeda A (1983) Structure of the envelope of Escherichia coli observed by the rapid-freezing and substitution fixation method. Microbiol Immunol 27:95.

Andrew SM, Jasani B (1987) An improved method for the inhibition of endogenous peroxidase non-deleterious to lymphocyte surface markers. Application to immunoperoxidase studies on eosinophil-rich tissue preparations. Histochem J 19:426.

Armbruster BL, Carlemalm E, Chiovetti R, Garavito RM, Hobot JA, Kellenberger E, Villiger W (1982) Specimen perparation for electron microscopy using lowtemperature embedding resins. J Microsc (Oxford) 126:77.

Armbruster BL, Kellenberger E, Carlemalm E, Villiger W, Garavito RM, Hobot JA, Chiovetti R, Acetarin JD (1984) Lowicryl resins - present and future applications. In: Revel JP, Barnard T, Haggis GH (eds) The Science of Biological Specimen Preparation for Microscopy and Microanalysis. SEM Inc., AMF O'Hare, Chicago, Il, pp 77-81.

Artymink PJ, Blake CCF, Grace DEP, Oatley SJ, Philipps DC, Sternberg MJE (1979) Crystallographic studies of the dynamic properties of lysozyme. Nature 280:563.

Ashford AE, Allaway WG, Gubler F, Lennon A, Sleegers J (1986) Temperature control in Lowicryl K4M and glycol methacrylate during polymerisation: is there a low temperature method? J Microsc (Oxford) 144:107.

Bachhuber K, Frösch D (1983) Melamine resins - a new class of water-soluble embedding media for electron microscopy. J Microsc (Oxford) 130:1.

Bachmann L, Schmitt WW (1971) Improved cryofixation applicable to freeze-etching. Proc Natl Acad Sci (USA) 68:2149.

Bald WB (1983) Optimizing the cooling block for the quick freeze method. J Microsc (Oxford) 131:11.

Bald WB (1985) The relative merits of various cooling methods. J Microsc (Oxford) 140:17.

Baschong W, Wrigley NG (1990) Small colloidal gold conjugated to Fab fragments or to immunoglobulin G as high resolution labels for electron microscopy: a technical overview. J Electron Microsc Techn 14:313.

Baschong W, Baschong-Prescianotto C, Wurtz M, Carlemalm E, Kellenberger C, Kellenberger E (1984) Preservation of protein structures for electron microscopy by fixation with aldehydes and/or OsO_4. Eur J Cell Biol 35:21.

Baskin DG, Erlandsen SL, Parsons JA (1979) Influence of hydrogen peroxide or alcholic sodium hydroxide on the immunocytochemical detection of growth hormone and prolactin after osmium fixation. J Histochem Cytochem 27:1290.

Bendayan M (1981) Ultrastructural localization of nucleic acids by the use of enzyme gold complexes. J Histochem Cytochem 29:531.

Bendayan M (1984) Protein A-gold electron microscopic immuncytochemistry: methods, applications and limitations. J Electron Microsc Techn 1:243.

Bendayan M (1987) Introduction of the protein G-gold complex for high-resolution immunocytochemistry. J Electron Microsc Techn 6:7.

Bendayan M, Garzon S (1988) Protein G-gold complex: comparative evaluation with protein A-gold for high-resolution immunocytochemistry. J Histochem Cytochem 36:597.

Bendayan M, Zollinger M (1983) Ultrastructural localization of antigenic sites on osmium-fixed tissues applying the protein A-gold technique. J Histochem Cytochem 31:101.

Bendayan M, Nanci A, Kan FWK (1987) Effect of tissue processing on colloidal gold cytochemistry. J Histochem Cytochem 35:983.

Bendayan M, Roth J, Perrelet A, Orci L (1980) Quantitative immunocytochemical localization of pancreatic secretory proteins in subcellular compartments of the rat acinar cell. J Histochem Cytochem 28:149.

Bendayan M, Barth RF, Gingras D, Londoño I, Robinson PT, Alam F, Adams DM, Mattiazzi L (1989) Electron spectroscopic imaging for high-resolution immunocytochemistry: use of boronated protein A. J Histochem Cytochem 37:573.

Berryman MA, Rodewald RD (1990) An enhanced method for post-embedding immunocytochemical staining which preserves cell membranes. J Histochem Cytochem 38:159.

Bigbee JW, Kosek JC, Eng LF (1977) Effects of primary antiserum dilution on staining 'antigen rich' tissues with the peroxidase antiperoxidase technique. J Histochem Cytochem 25:443.

Björck L, Kronvall G (1984) Purification and some properties of Streptococcal protein G, a novel IgG-binding reagent. J Immunol 133:969.

Björnsti MA, Hobot JA, Kelus AS, Villiger W, Kellenberger E (1986) New electron microscopic data on the structure of the nucleoid and their functional consequences. In: Gualerzi CO, Pons CL (eds) Bacterial Chromatin. Springer-Verlag, Berlin, pp 64-81.

Bonnard C, Papermaster DS, Kraehenbuhl JP (1984) The streptavidin-biotin bridge technique: application in light and electron microscope immuncytochemistry. In: Polak JS, Varndell IM (eds) Immunolabelling for Electron Microscopy. Elsevier, Amsterdam, pp 95-111.

Bowdler AL (1991) An assessement of the electron beam stable, acrylic resin, LR White as a single embedding medium for the diagnostic analysis of renal biopsy sections by light and electron microscopy. Thesis for the Fellowship of the Institute of Medical Laboratory Sciences, London.

Bowdler AL, Griffiths DFR, Newman GR (1989) The morphological and immunohistochemical analysis of renal biopsies by light and electron microscopy using a single processing method. Histochem J 21:393.

Carlemalm E, Kellenberger E (1982) The reproducible observation of unstained embedded cellular material in thin sections: visualisation of an integral membrane protein by a new mode of imaging for STEM. EMBO J 1:63.

Carlemalm E, Villiger W (1989) Low temperature embedding. In: Bullock GR, Petrusz P (eds) Techniques for Immunocytochemistry, Vol.4. Academic Press, London, pp 29-45.

Carlemalm E, Coliex C, Kellenberger E (1985a) Contrast formation in electron microscopy of biological material. Adv Electron Electron Phys 63:269.

Carlemalm E, Garavito RM, Villiger W (1980) Advances in low temperature embedding for electron microscopy. In: Brederoo P, de Priester W (eds) Proceedings of the Seventh European Congress on Electron Microscopy, The Hague, The Netherlands, Vol.2, pp 656-657.

Carlemalm E, Garavito RM, Villiger W (1982a) Resin development for electron microscopy and an analysis of embedding at low temperature. J Microsc (Oxford) 126:123.

Carlemalm E, Villiger W, Hobot JA, Acetarin J-D, Kellenberger E (1985b) Low temperature embedding with Lowicryl resins: two new formulations and some applications. J Microsc (Oxford) 140:55.

Carlemalm E, Armbruster BL, Chiovetti R, Garavito RM, Hobot JA, Villiger W, Kellenberger E (1982b) Perspectives for achieving improved information by the observation of thin sections in the electron microscope. Tokai J Exp Clin Med 7 (Suppl):33.

Causton BE (1980) The molecular structure of resins and its effect on the epoxy embedding resins. Proc Roy Microsc Soc 15:185.

Causton BE (1984) The choice of resin for electron immunocytochemistry. In: Polak JM, Varndell IM (eds) Immunolabelling for Electron Microscopy. Elsevier, Amsterdam, pp 29-36

Causton BE, Ashurst DE, Butcher, RG, Chapman SK, Thomson DJ, Webb MJW (1981) Resins: toxicity and safe handling. Proc Roy Microsc Soc 16:265.

Childs GV, Unabia G (1982) Application of a rapid avidin-biotin-peroxidase complex (ABC) technique to the localization of pituitary hormones at the electron microscopic level. J Histochem Cytochem 30:1320.

Chiovetti R, Little SA, Brass-Dale J, McGuffee LJ (1986) A new approach to low temperature embedding: quick freezing, freeze-drying and direct infiltration in Lowicryl K4M. In: Müller M, Becker RP, Boyde A, Wolosewick JJ (eds) The Science of Biological Specimen Perparation for Microscopy and Microanalysis. Scanning Electron Microscopy Inc., AMF O'Hare, Chicago, Illinois, pp 155-164.

Chiovetti R, McGuffee LJ, Little SA, Wheeler-Clark E, Brass-Dale J (1987) Combined quick freezing, freeze-drying, and embedding tissue at low temperature and in low viscosity resin. J Electron Microsc Techn 5:1.

Clarke-Sturman AJ, Archibald AR, Hancock IC, Harwood CR, Merad T, Hobot JA (1989) Cell wall assembly in Bacillus subtilis: partial conservation of polar wall material and the effect of growth conditions on the pattern of incorporation of new material at the polar caps. J Gen Microbiol 135:657.

Coggi G, Dell'Orto P, Grigalato PG, Sacchi G, Vialli G (1984) Immunoelectron microscopy of human renal biopsies: pre-requisites and limitations. Appl Pathol 2:223.

Coons AH, Creech HJ, Jones RN (1941) Immunological properties of an antibody containing a fluorescent group. Proc Soc Exp Biol 47:200.

Craig S, Miller C (1984) LR White resin and improved on-grid immunogold detection of vicilin, a pea seed storage protein. Cell Biol Int Rep 8:879.

Dahmen H, Hobot JA (1986) Ultrastructural analysis of Erysiphe graminis haustoria and subcuticular stroma of Venturia inaequalis using cryosubstitution. Protoplasma 131:92.

Danscher G, Rytter Norgaard JO (1983) Light microscopic visualisation of colloidal gold on resin-embedded tissue. J Histochem Cytochem 31:1394.

De Mey J (1983) Colloidal gold probes in immunocytochemistry. In: Polak JM, Van Noorden S (eds) Immunocytochemistry - Practical Applications in Pathology and Biology. Wright PSG, Boston, pp 82-112.

Douzou P (1977) Enzymology at sub-zero temperatures. Adv Enzymol 45:157.

Douzou P, Hoa GHB, Petsko GA (1975) Protein crystallography at sub-zero temperatures: lysozyme-substrate complexes in cooled mixed solvents. J Mol Biol 96:367.

Douzou P, Hoa GHB, Maurel P, Travers F (1976) Physical chemical data for mixed solvents used in low temperature biochemistry. In: Fasman GD (ed) Handbook of Biochemistry and Molecular Biology, Vol.1, 3rd. edition. CRC Press Inc., Cleveland, Ohio, pp 520-539.

Dubochet J, Lepault J, Freeman R, Berriman JA, Homo J-C (1982) Electron microscopy of frozen water and aqueous solutions. J Microsc (Oxford) 128:219.

Dubochet J, McDowall AW, Menge B, Schmid BN, Lickfeld KG (1983) Electron microscopy of frozen-hydrated bacteria. J Bacteriol 155:381.

Dudek RW, Boyne AF, Freinkel N (1981) Quick-freeze fixation and freeze-drying of isolated rat pancreatic islets: application to the ultrastructural localization of inorganic phosphate in the pancreatic beta cell. J Histochem Cytochem 29:321.

Dudek RW, Childs GV, Boyne AF (1982) Quick-freezing and freeze-drying in preparation for high quality morphology and immunocytochemistry at the ultrastructural level: application to pancreatic beta cell. J Histochem Cytochem 30:129.

Dudek RW, Varndell IM, Polak JM (1984) Combined quick-freeze and freeze-drying techniques for improved electron immunocytochemistry. In: Polak JS, Varndell IM (eds) Immunolabelling for Electron Microscopy. Elsevier, Amsterdam, pp 235-248.

Dürrenberger M (1989) Removal of background label in immunocytochemistry with the apolar Lowicryls by using washed protein A-gold precoupled antibodies in a one-step procedure. J Elctron Microsc Techn 11:109.

Dürrenberger M, Björnsti MA, Uetz T, Hobot JA, Kellenberger E (1988) Intracellular location of the histone-like protein HU in Escherichia coli. J Bacteriol 170:4757.

Ebersold HR, Cordier JL, Lüthy P (1981) Bacterial mesosomes: method dependent artifacts. Arch Microbiol 130:19.

Edelmann L (1986) Freeze-dried embedded specimens for microanalysis. In: Johari O (ed) Scanning Electron Microscopy, Part IV. Scanning Electron Microscopy Inc., AMF O'Hare, Chicago, Illinois, pp 1337-1356.

Edelmann L (1989a) The contracting muscle: a challenge for freeze-substitution and low temperature embedding. In: Albrecht RM, Ornberg RL (eds) Scanning Microscopy Supplement 3. Scanning Microscopy Intl., AMF O'Hare, Chicago, Illinois, pp 241-252.

Edelmann L (1989b) Freeze substitution and low temperature embedding. Inst Phys Conf Ser 98:763.

Edelmann L (1991) Freeze-substitution and the preservation of diffusable ions. J Microsc (Oxford) 161:217.

Eliasson M, Anderson R, Olsson A, Wigzell H, Uhlén M (1989) Differential IgG-binding characteristics of Staphylococcal protein A, Streptococcal protein G, and a chimeric protein AG. J Immunol 142:575.

Eliasson M, Olsson A, Palmcrantz E, Wiberg K, Inganäs M, Guss B, Lindberg M, Uhlén M (1988) Chimeric IgG-binding receptors engineered from Staphylococcal protein A and Streptococcal protein G. J Biol Chem 263:4323.

Ellinger A, Pavelka M (1985) Post-embedding localisation of glycoconjugates by means of lectins on thin sections of tissues embedded in LR White. Histochem J 17:1321.

Emerman M, Behrman EJ (1982) Cleavage and crosslinking of proteins with osmium(VIII) reagents. J Histochem Cytochem 30:395.

Epstein W, Schultz SG (1965) Cation transport in Escherichia coli. V. Regulation of cation content. J Gen Physiol 49:221.

Erickson PA, Anderson DH, Fisher SK (1987) Use of uranyl acetate en bloc to improve tissue preservation and labeling for post-embedding immunoelectron microscopy. J Electron Microsc Techn 5:303.

Escaig J (1982) New instruments which facilitate freezing at 83 K and 6 K. J Microsc (Oxford) 126:221.

Faraday M (1857) Experimental relations of gold (and other metals) to light. Philos Trans Roy Soc London 147:145.

Faulk WP, Taylor GM (1971) An immunocolloid method for the electron microscope. Immunochemistry 8:1081.

Feldherr CM, Marshall JM (1962) The use of colloidal gold for studies of intracellular exchange in amoeba Chaos chaos. J Cell Biol 12:640.

Fernández-Morán H (1960) Low temperature preparation techniques for electron microscopy of biological specimens based on rapid freezing with liquid helium II. Ann N Y Acad Sci 85:689.

Fernández-Morán H (1961) The fine structure of vertebrate and invertebrate photoreceptors as revealed by low temperature microscopy. In: Smelser GK (ed) The Structure of the Eye. Academic Press, New York, pp 521-556.

Fink AL, Ahmed IA (1976) Formation of stable crystalline enzyme-substrate intermediates at sub-zero temperatures. Nature 263:294.

Frauenfelder H, Petsko GA, Tsernoglou D (1979) Temperature dependent X-ray diffraction as a probe of protein structural dynamics. Nature 280:558.

Frösch D, Westphal C (1985) Choosing the appropriate section thickness in the melamine embedding technique. J Microsc (Oxford) 137:177.

Frösch D, Westphal C (1989) Melamine resins and their application in electron microscopy. Electron Microsc Rev 2:231.

Fujiwara K, Yasuno M, Kitagawa T (1981) Novel preparation method of immunogen for hydrophobic hapten, enzyme immunoassay for daunomycin and adriamycin. J Immunol Meth 45:195.

Gallyas F, Gorcs T, Merchenthaler I (1982) High-grade intensification of the end-product of the diaminobenzidine reaction for peroxidase histochemistry. J Histochem Cytochem 30:183.

Garavito RM, Carlemalm E, Colliex C, Villiger W (1982) Septate junction ultrastructure as visualised in unstained and stained preparations. J Ultrastruct Res 80:344.

Gee JMW, Nicholson RI, Jasani B, Newman GR, Amselgruber WM (1990) An immunocytochemical method for localization of estrogen receptors in rat tissues using a dinitrophenyl (DNP)-labeled rat monoclonal primary antibody. J Histochem Cytochem 38:69.

Geoghegan WD (1988) The effect of three variables on adsorption of rabbit IgG to colloidal gold. J Histochem Cytochem 36:401.

Geoghegan WD (1991) An electrophoretic method for selection of conditions for production of electrophoretically uniform protein coloidal gold complexes. J Histochem Cytochem 39:111.

Ghitescu L, Bendayan M (1990) Immunolabeling efficiency of protein A-gold complexes. J Histochem Cytochem 38:1523.

Ghitescu L, Galis Z, Bendayan M (1991) Protein AG-gold complex: an alternative probe in immunocytochemistry. J Histochem Cytochem 39:1057.

Gibbons IR (1959) An embedding resin miscible with water for electron microscopy. Nature (London) 184:375.

Gillett R, Gull K (1972) Glutaraldehyde - its purity and stability. Histochemie 30:162.

Glauert AM (1965) The fixation and embedding of biological specimens. In: Kay D (ed) Techniques for Electron Microscopy, 2nd. edition. Blackwell Scientific Publications, Oxford, pp 166-212.

Glauert AM (1975) Fixation, Dehydration, and Embedding of Biological Specimens. Practical Methods in Electron Microscopy, Vol.3, Part I. Elsevier/North Holland, Amsterdam.

Glauert AM (1991) Epoxy resins: an update on their selection and use. Microscopy and Analysis 25:15.

Glauert AM, Glauert RH (1958) Araldite as an embedding medium for electron microscopy. J Biophys Biochem Cytol 4:191.

Glauert AM, Young RD (1989) The control of temperature during polymerisation of Lowicryl K4M: There is a low temperature method. J Microsc (Oxford) 154:101.

Glauert AM, Rogers GE, Glauert RH (1956) A new embedding medium for electron microscopy. Nature (London) 178:803.

Graber MB, Kreutzberg GW (1985) Immunogold staining (IGS) for electron microscopal demonstration of glial fibrillary acidic (GFA) protein in LR White embedded tissue. Histochemistry 83:497.

Griffiths G, Hoppeler H (1986) Quantitation in immunocytochemistry: correlation of immunogold labeling to absolute number of membrane antigens. J Histochem Cytochem 34:1389.

Handley DA (1989a) The development and application of colloidal gold as a microscopic probe. In: Hayat MA (ed) Colloidal Gold: Principles, Methods, and Applications, Vol.1. Academic Press, San Diego, pp 1-12.

Handley DA (1989b) Methods of synthesis of colloidal gold. In: Hayat MA (ed) Colloidal Gold: Principles, Methods, and Applications, Vol.1. Academic Press, San Diego, pp 13-32.

Hayat MA (1981) Fixation for Electron Microscopy. Academic Press, New York.

Hayat MA (1986) Glutaraldehyde: role in electron microscopy. Micron Microsc Acta 17:115.

Hayat MA (1989) Principles and Techniques of Electron Microscopy: Biological Applications, 3rd. edition. Macmillan, London, and CRC Press, Boca Raton, Florida.

Hermann R, Müller M (1991) High resolution biological scanning electron microscopy: a comparative study of low temperature metal coating techniques. J Electron Microsc Techn 18:440.

Hernández Mariné MC (1992) A simple way to encapsulate small samples for processing for TEM. J Microsc (Oxford) 168:203.

Heuser JE, Reese TS, Dennis MJ, Jan, Y, Jan L, Evans L (1979) Synaptic vesicle exocytosis captured by quick freezing and correlated with quantal transmitter release. J Cell Biol 81:275.

Hewlins MJE, Weeks I, Jasani B (1984) Non-deleterious dinitrophenyl (DNP) hapten labelling of antibody protein. Preparation and properties of some short-chain DNP imidoesters. J Immunol Methods 70:111.

Hippe S, Hermanns M (1986) Improved structural preservation in freeze-substituted sporidia of Ustilago avenae - a comparison with low temperature embedding. Protoplasma 135:19.

Hittmair A, Schmid KW (1989) Inhibition of endogenous peroxidase for the immunocytochemical demonstration of intermediate filament proteins (IFP). J Immunol Methods 116:199.

Hobot JA (1989) The Lowicryls and low temperature embedding for colloidal gold methods. In: Hayat MA (ed) Colloidal Gold: Principles, Methods, and Applications, Vol.2. Academic Press, San Diego, pp 75-115.

Hobot JA (1990) New aspects of bacterial ultrastructure as revealed by modern acrylics for electron microscopy. J Struct Biol 104:169.

Hobot JA (1991) Low temperature embedding techniques for studying microbial cell surfaces. In: Mozes N, Handley P, Busscher HJ, Rouxhet PG (eds) Microbial Cell Surface Analysis: Structural and Physico-Chemical Methods. VCH Publishers, New York, pp 127-150.

Hobot JA, Newman GR (1990) Electron microscopy for bacterial cells. In: Harwood CR, Cutting SM (eds) Molecular Biology Methods for Bacillus. John Wiley and Sons, Chichester, pp 352-362.

Hobot JA, Newman GR (1991) Strategies for improving the cytochemical and immunocytochemical sensitvity of ultrastructurally well-preserved, resin embedded biological tissue for light and electron microscopy. In: Roomans G, Edelmann L (eds) Scanning Microscopy Supplement 5. Scanning Microscopy Intl., AMF O'Hare, Chicago, Illinois, pp S27-S41.

Hobot JA, Rogers HJ (1991) Intracellular localization of the autolytic N-acetylmuramyl-L-alanine amidase in Bacillus subtilis 168 and in an autolysis-deficient mutant by immunoelectron microscopy. J Bacteriol 173:961.

Hobot JA, Björnsti MA, Kellenberger E (1987) Use of on-section immunolabeling and cryosubstitution for studies of bacterial DNA distribution. J Bacteriol 169:2055.

Hobot JA, Carlemalm E, Kellenberger E (1981) High resolution electron microscopy of stained vs unstained bacterial cell envelopes in CTEM and STEM. Experientia 37:1226.

Hobot JA, Felix HR, Kellenberger E (1982) Ultrastructure of permeabilised cells of Escherichia coli and Cephalosporium acremonium. FEMS Microbiol Lett 13:57.

Hobot JA, Carlemalm E, Villiger W, Kellenberger E (1984) Periplasmic gel: new concept resulting from the reinvestigation of bacterial cell envelope ultrastructure by new methods. J Bacteriol 160:143.

Hobot JA, Villiger W, Escaig J, Maeder M, Ryter A, Kellenberger E (1985) Shape and fine structure of nucleoids observed on sections of ultra-rapidly frozen and cryosubstituted bacteria. J Bacteriol 162:960.

Hoch HC, Howard RJ (1980) Ultrastructure of freeze-substituted hyphae of the basidiomycete Laetisaria arvalis. Protoplasma 103:281.

Holgate CS, Jackson P, Cowen PN, Bird CC (1983) Immunogold-silver staining. New method of immunostaining with enhanced sensitivity. J Histochem Cytochem 31:938.

Hopwood D, Coghill G, Ramsey J, Milne G, Kerr M (1984) Microwave fixation: its potential for routine techniques, histochemistry, immunocytochemistry and electron microscopy. Histochem J 16:1171.

Horisberger M, Clerc M-F (1985) Labelling of colloidal gold with protein A. A quantitative study. Histochemistry 82:219.

Horowitz RA, Giannasca PJ, Woodcock CL (1990) Ultrastructural preservation of nuclei and chromatin: improvement with low temperature methods. J Microsc (Oxford) 157:205.

Howard RJ (1981) Ultrastructural analysis of hyphal tip growth in fungi: Spitzenkorper, cytoskeleton and endomembranes after freeze-substitution. J Cell Sci 48:89.

Hsu S-M, Raine L, Fanger H (1981) Use of avidin-biotin-peroxidase complex (ABC) in immunoperoxidase techniques: a comparison between ABC and unlabelled antibody (PAP) procedures. J Histochem Cytochem 29:577.

Humbel BM (1984) Gefriersubstiution - Ein Weg zur Verbesserung der morphologischen und zytologischen Untersuchung biologischer Proben im Elektronenmikroskop. Ph.D Thesis (7609), ETH, Zürich.

Humbel B, Müller M (1984) Freeze substitution and low temperature embedding. Proc 8th Eur Congr Electron Microsc 3:1789.

Humbel B, Marti T, Müller M (1983) Improved structural preservation by combining freeze-substitution and low temperature embedding. Beitr Elektronmikrosk Direktabb Oberfl 16:585.

Hunziker EB, Herrmann W (1987) In situ localization of cartilage extracellular matrix components by immunoelectron microscopy after cytochemical tissue processing. J Histochem Cytochem 35:647.

Idelman S (1964) Modification de la technique de Luft en vue de la conservation des lipides en microscopie électronique. J Microscopie 3:715.

Idelman S (1965) Conservation des lipides par les techniques utilisées en microscopie électronique. Histochemie 5:18.

Ito S, Winchester RJ (1963) The fine structure of the gastric mucosa in the bat. J Cell Biol 16:541.

Jasani B, Williams ED (1980) DNP as a hapten in immunolocalisation studies. J Med Microbiol 13:xv.

Jasani B, Wynford-Thomas D, Williams ED (1981) Use of monoclonal anti-hapten antibodies for immunolocalisation of tissue antigens. J Clin Pathol 34:1000.

Jasani B, Wynford-Thomas D, Thomas NB, Newman GR (1986) Broad-spectrum, non-deleterious inhibition of endogenous peroxidase: LM and EM application. Histochem J 18:56.

Jesaitis AJ, Buescher ES, Harrison D, Quinn MT, Parkos CA, Livesey S, Linner J (1990) Ultrastructural localization of cytochrome b in the membranes of resting and phagocytosing human granulocytes. J Clin Invest 85:821.

Jones CJP, Stoddart RW (1986) A post-embedding avidin-biotin peroxidase system to demonstrate the light and electron microscopic localisation of lectin binding sites in rat tubules. Histochem J 18:371.

Jones JT, Gwynn IA (1991) A method for rapid fixation and dehydration of nematode tissue for transmission electron microscopy. J Microsc (Oxford) 164:43.

Jorgensen AO, McGuffee LJ (1987) Immunoelectron microscopic localisation of sarcoplasmic reticulum proteins in cryofixed, freeze-dried, and low temperature embedded tissue. J Histochem Cytochem 35:723.

Kandasamy MK, Parthasarathy MV, Nasrallah ME (1991) High pressure freezing and freeze substitution improve immunolabeling of S-locus specific glycoproteins in the stigma papillae of Brassica. Protoplasma 162:187.

Kellenberger E (1991) The potential of cryofixation and freeze substitution: observations and theoretical considerations. J Microsc (Oxford) 161:183.

Kellenberger E, Hayat MA (1991) Some basic concepts for the choice of methods. In: Hayat MA (ed) Colloidal Gold: Principles, Methods, and Applications, Vol.3. Academic Press, San Diego, pp 1-30.

Kellenberger E, Ryter A (1964) In bacteriology. In: Siegel BM (ed) Modern Developments in Electron Microscopy. Academic Press, London, pp 335-393.

Kellenberger E, Séchaud J, Blondel B (1972) Appendix in Séchaud J, Kellenberger E (1972) Electron microscopy of DNA-containing plasms. IV. J Ultrastruct Res 39:598.

Kellenberger E, Carlemalm E, Villiger W, Roth J, Garavito RM (1980) Low Denaturation Embedding for Electron Microscopy of Thin Sections. Chemische Werke Löwi GmbH, P.O. Box 1660, D-8264 Waldkraiburg, Germany.

Kellenberger E, Dürrenberger M, Villiger W, Carlemalm, E, Wurtz M (1987) The efficiency of immunolabel on Lowicryl sections compared to theoretical predictions. J Histochem Cytochem 35:959.

Knoll G, Oebel G, Plattner H (1982) A simple sandwich-cryogen-jet procedure with high cooling rates for cryofixation of biological materials in the native state. Protoplasma 111:161.

Kraehenbuhl JP, Racine L, Jamieson JD (1977) Immunocytochemical localization of secretory proteins in bovine pancreatic exocrine cells. J Cell Biol 72:406.

Kramarcy NR, Sealock R (1991) Commercial preparations of colloidal gold-antibody complexes frequently contain free active antibody. J Histochem Cytochem 39:37.

Kushida H (1961a) A styrene-methacrylate resin embedding method for ultrathin sectioning. J Electron Microscopy 10:16.

Kushida H (1961b) A new embedding method for ultrathin sectioning using a methacrylate resin with three dimensional polymer structure. J Electron Microscopy 10:194.

Kushida H (1963) A modification of the water-miscible epoxy resin 'Durcupan' embedding method for ultrathin sectioning. J Electron Microscopy 12:71.

Kushida H (1964) Improved methods for embedding with Durcupan. J Electron Microscopy 13:139.

Kushida H (1966) Further improved method for embedding with Durcupan. J Electron Microscopy 15:95.

Kushida H (1971) A new method for embedding with Epon 812. J Electron Microscopy 20:206.

Leduc EH, Bernhard W (1962) Water-soluble embedding media for ultrastructural cytochemistry. Digestion with nucleases and proteinases. In: Harris RJC (ed) The Interpretation of Ultrastructure. Academic Press, New York, pp 21-45.

Lin N, Langenberg WG (1983) Immunohistochemical localisation of barley stripe mosaic virions in infected wheat cells. J Ultrastruct Res 84:16.

Lin N, Langenberg WG (1984) Chronology of appearance of barley stripe virus protein in infected wheat cells. J Ultrastruct Res 89:309.

Linner JG, Livesey SA (1988) Low temperature molecular distillation drying of cryofixed biological samples. In: McGrath JJ, Diller KR (eds) Low Temperature Biotechnology: Emerging Applications and Engineering Contributions, Vol.10. American Society of Mechanical Engineers, New York, pp 147-157.

Linner JG, Livesey SA, Harrison DS, Steiner AL (1986) A new technique for removal of amorphous phase tissue water without ice crystal damage: a preparative method for ultrastructural analysis and immunoelectron microscopy. J Histochem Cytochem 34:1123.

Livesey SA, del Campo AA, McDowall AW, Stasny JT (1991) Cryofixation and ultra-low-temperature freeze-drying as a preparative technique for TEM. J Microsc (Oxford) 161:205.

Livesey SA, Buescher ES, Krannig GL, Harrison DS, Linner JG, Chiovetti R (1989) Human neutrophil granule heterogeneity: immunolocalization studies using cryofixed, dried and embedded specimens. In: Albrecht RM, Ornberg RL (eds) Scanning Microscopy Supplement 3. Scanning Microscopy Intl., AMF O'Hare, Chicago, Illinois, pp 231-240.

Londoño I, Coulombe PA, Bendayan MA (1989) Brief review on progress in enzyme-gold cytochemistry. In: Albrecht RM, Ornberg RL (eds) Scanning Microscopy Supplement 3. Scanning Microscopy Intl., AMF O'Hare, Chicago, Illinois, pp 7-14.

Lucocq J (1992) Quantitation of gold labeling and estimation of labeling efficiency with a stereological counting method. J Histochem Cytochem 40:1929.

Luft JH (1961) Improvements in epoxy resin embedding methods. J Biophys Biochem Cytol 9:409.

Maaløe O, Birch-Andersen A (1956) On the organization of the 'nuclear material' in Salmonella typhimurium. Symp Soc Gen Microbiol 6:261.

MacKenzie AP (1972) Freezing, freeze-drying and freeze-substitution. In: Johari O (ed) Scanning Electron Microscopy, Vol.2. Scanning Electron Microscopy Inc., AMF O'Hare, Chicago, Illinois, pp 273-280.

Mar H, Wight TN (1988) Colloidal gold immunostaining on deplasticised ultra-thin sections. J Histochem Cytochem 36:1387.

McLean IW, Nakane PK (1974) Periodate-lysine-paraformaldehyde fixative, a new fixative for immunoelectron microscopy. J Histochem Cytochem 22:1077.

Menco BPM (1986) A survey of ultra-rapid cryofixation methods with particular emphasis on applications to freeze-fracturing, freeze-etching and freeze-substitution. J Electron Microsc Techn 4:177.

Merad T, Archibald AR, Hancock IC, Harwood CR, Hobot JA (1989) Cell wall assembly in Bacillus subtilis: visualization of old and new wall material by electron microscopic examination of samples stained selectively for teichoic acid and teichuronic acid. J Gen Microbiol 135:645.

Millonig G (1961) A modified procedure for lead staining of thin sections. J Biophys Biochem Cytol 11:736.

Monaghan P, Robertson D (1990) Freeze-substitution without aldehyde or osmium fixatives: ultrastructure and implications for immunocytochemistry. J Microsc (Oxford) 158:355.

Moncany MLJ (1982) Détermination des conditions intracellulaires chez E. coli. Conséquences biologiques de leur modification. Thesis Docteur d'Etat, Université de Paris VII, Paris.

Moor H (1987) Theory and practice of high pressure freezing. In: Steinbrecht RA, Zierold K (eds) Cryotechniques in Biological Electron Microscopy. Springer, Berlin, pp 175-191.

Müller M, Meister N, Moor H (1980) Freezing in a propane jet and its application in freeze-fracturing. Mikroskopie (Wien) 36:129.

Mutasa HCF (1989) Applicability of using acrylic resins in post-embedding ultrastructural immunolabeling of human neutrophil granule proteins. Histochem J 21:249.

Nakane PK, Pierce GB (1966) Enzymes-labeled antibodies: preparation and application for the localization of antigens. J Histochem Cytochem 14:929.

Neiss WF (1988) Enhancement of the periodic acid-Schiff (PAS) and periodic acid thiocarbohydrazide-silver proteinate (PA-TCH-SP) reaction in LR White sections. Histochemistry 88:603.

Newman GR (1987) Use and abuse of LR White. Histochem J 19:118.

Newman GR (1989) LR White embedding medium for colloidal gold methods. In: Hayat MA (ed) Colloidal Gold: Principles, Methods, and Applications, Vol.2. Academic Press, San Diego, pp 47-73.

Newman GR, Hobot JA (1987) Modern acrylics for post-embedding immunostaining techniques. J Histochem Cytochem 35:971.

Newman GR, Hobot JA (1989) Role of tissue processing in colloidal gold methods. In: Hayat MA (ed) Colloidal Gold: Principles, Methods, and Applications, Vol.2. Academic Press, San Diego, pp 33-45.

Newman GR, Jasani B (1984a) Post-embedding immunoenzyme techniques. In: Polak JS, Varndell IM (eds) Immunolabelling for Electron Microscopy. Elsevier, Amsterdam, pp 53-70.

Newman GR, Jasani B (1984b) Immunoelectron microscopy: immunogold and immunoperoxidase compared using a new post-embedding system. Med Lab Sci 41:238.

Newman GR, Jasani B, Williams ED (1982) The preservation of ultrastructure and antigenicity. J Microsc (Oxford) 127:RP5-RP6.

Newman GR, Jasani B, Williams ED (1983a) A simple post-embedding system for the rapid demonstration of tissue antigens under the electron microscope. Histochem J 15:543.

Newman GR, Jasani B, Williams ED (1983b) Metal compound intensification of the electron density of diaminobenzidine. J Histochem Cytochem 31:1430.

Newman GR, Jasani B, Williams ED. (1983c) The visualisation of trace amounts of diaminobenzidine (DAB) polymer by a novel gold-sulphide-silver method. J Microsc (Oxford) 132:RP1-RP2.

Newman GR, Jasani B, Williams ED (1986) Multiple hormone storage by polycrine cells in the pancreas (from a case of nesidioblastosis). Histochem J 18:67.

Newman GR, Jasani B, Williams ED (1989) Multiple hormone storage by cells of the human pituitary. J Histochem Cytochem 37:1183.

Nicolas G (1991) Advantages of fast-freeze fixation followed by freeze-substitution for the preservation of cell integrity. J Electron Microsc Techn 18:395.

Pappas PW (1971) The use of a chrome alum-gelatin (subbing) solution as a general adhesive for paraffin sections. Stain Technology 46:121.

Pease DC (1964) Histological Techniques for Electron Microscopy, 2nd. edition. Academic Press, New York.

Pelletier G, Morel G (1984) Immunoenzyme techniques at the electron microscopical level. In: Polak JS, Varndell IM (eds) Immunolabelng for Electron Microscopy. Elsevier, Amsterdam, pp 83-93.

Plattner H, Bachmann L (1982) Cryofixation: a tool in biological ultrastructural research. Int Rev Cytol 79:237.

Plattner H, Knoll G (1984) Cryofixation of biological materials for electron microscopy by the methods of spray-, sandwich-, cryogen-jet- and sandwich-jet-freezing: a comparison of techniques. In: Revel JP, Barnard T, Haggis GH (eds) Science of Biological Specimen Preparation for Microscopy and Microanalysis. Scanning Electron Microscopy Inc., AMF O'Hare, Chicago, Il, pp 139-146.

Posthuma G, Slot JW, Geuze HJ (1984) Immunocytochemical assay of amylase and chymotrypsinogen in rat pancreas secretory granules. J Histochem Cytochem 32:1028.

Posthuma G, Slot JW, Geuze HJ (1986) A quantitative immunoelectron microscopic study of amylase and chymotrypsinogen in peri- and tele-insular cells of the rat exocrine pancreas. J Histochem Cytochem 34:203.

Posthuma G, Slot JW, Geuze HJ (1987) Usefulness of the immunogold technique in quantitation of a soluble protein in ultra-thin sections. J Histochem Cytochem 35:405.

Quintana C (1993) In situ conservation of diffusable elements in liver cells after cryofixation, cryosubstitution, and low temperature embedding at 193K in HM23 Lowicryl resin. Microsc Res Techn 24:103.

Reid N, Beesley JE (1991) Sectioning and Cryosectioning for Electron Microscopy. Practical Methods in Electron Microscopy, Vol.13. Glauert AM (ed). Elsevier, Amsterdam.

Reid N (1975) Ultramicrotomy. Practical Methods in Electron Microscopy, Vol.3, Part II. Glauert AM (ed). Elsevier, Amsterdam.

Reynolds ES (1963) The use of lead citrate at high pH as an electron opaque stain in the electron microscope. J Cell Biol 17:208.

Robards AW, Sleytr UB (1985) Low Temperature Methods in Biological Electron Microscopy. Practical Methods in Electron Microscopy, Vol.10. Glauert, A.M. (ed.). Elsevier, Amsterdam.

Romano EL, Romano M (1977) Staphylococcal protein A bound to colloidal gold: a useful reagent to label antigen-antibody sites in electron microscopy. Immunochemistry 14:711.

Roos N, Kinde U, Morgan JA (1990) Morphology of rat exocrine pancreas prepared by anhydrous cryo-procedures. J Electron Microsc Techn 14:39.

Rosenberg M, Bartl P, Lesko J (1960) Water-soluble methacrylate as an embedding medium for the preparation of ultrathin sections. J Ultrastruct Res 4:298.

Roth J (1983) Application of lectin gold complexes for electron microscopic localisation of glycoconjugates on thin sections. J Histochem Cytochem 31:987.

Roth J, Bendayan M, Orci L (1978) Ultrastructural localization of intracellular antigens by the use of protein A-gold complex. J Histochem Cytochem 26:1074.

Roth J, Bendayan M, Carlemalm E, Villiger W, Garavito RM (1981) Enhancement of structural preservation and immunocytochemical staining in low temperature embedded pancreatic tissue. J Histochem Cytochem 29:663.

Roth J, Taatjees DJ, Lucocq JN, Weinstein J, Paulson JC (1985) Demonstration of an extensive transtubular network continuous with the Golgi apparatus stack that may function in glycosylation. Cell (Cambridge, Mass) 43:287.

Ryan KP (1992) Cryofixation of tissues for electron microscopy: a review of plunge cooling methods. Scann Microsc 6:715.

Ryan KP, Liddicoat MI (1987) Safety considerations regarding the use of propane and other liquefied gases as coolants for rapid freezing purposes. J Microsc (Oxford) 147:337.

Ryter A, Kellenberger E, Birch-Anderson A, Maaløe O (1958) Étude au microscope électronique de plasmas contenant de l'acide désoxyribonucléique. Z Naturforscg 13b:597.

Sabatini DD, Bensch K, Barrnett RJ (1963) Cytochemistry and electron microscopy. The preservation of cellular ultrastructure and enzymic activity by aldehyde fixation. J Cell Biol 17:19.

Sandoz D, Nicolas G, Lainé MC (1985) Two mucous cell types revisted after quick-freezing and cryosubstitution. Biol Cell 54:79.

Schmid KW, Hittmair A, Schmidhammer H, Jasani B (1989) Non-deleterious inhibition of endogenous peroxidase activity (EPA) by cyclopropanone hydrate: a definitive approach. J Histochem Cytochem 37:473.

Schwarz H, Humbel BM (1989) Influence of fixatives and embedding media on immunolabeling of freeze-substituted cells. In: Albrecht RM, Ornberg RL (eds) Scanning Microscopy Supplement 3. Scanning Microscopy Intl., AMF O'Hare, Chicago, Illinois, pp 57-64.

Seligman AM, Karnovsky MJ, Wasserkrug HL, Hanker JS (1968) Non-droplet ultrastructural demonstration of cytochrome oxidase activity with a polymerizing osmiophilic reagent, diaminobenzidine (DAB). J Cell Biol 38:1.

Shinagawa Y (1972) Water-containing melamine-glutaraldehyde resin embedding method. J Electron Microsc 21:252.

Shinagawa Y, Shinagawa Y (1978) Melamine resin as water-containing embedding medium for electron microscopy. J Electron Microsc 27:13.

Simon GT, Thomas JA, Chorneyko KA, Carlemalm E (1987) Rapid embedding in Lowicryl K4M for immunoelectron microscopic studies. J Electron Microsc Techn 6:317.

Singer SJ (1959) Preparation of an electron dense antibody conjugate. Nature 183:1523.

Sitte H, Neumann K, Edelmann L (1986) Cryofixation and cryosubstitution for routine work in transmission electron microscopy. In: Müller M, Becker RP, Boyde A, Wolosewick JJ (eds) The Science of Biological Specimen Perparation for Microscopy and Microanalysis. Scanning Electron Microscopy Inc., AMF O'Hare, Chicago, Illinois, pp 103-118.

Sitte H, Edelmann L, Neumann K (1987a) Cryofixation without pretreatment at ambient pressure. In: Steinbrecht RA, Zierold K (eds) Cryotechniques in Biological Electron Microscopy, Springer-Verlag, Berlin, pp 87-113.

Sitte H, Neumann K, Edelmann L (1987b) Safety rules for cryopreparation. In: Steinbrecht RA, Zierold K (eds) Cryotechniques in Biological Electron Microscopy. Springer-Verlag, Berlin, pp 285-289.

Slot JW, Geuze HJ (1981) Sizing of protein A-colloidal gold probes for immunoelectron microscopy. J Cell Biol 90:533.

Slot JW, Geuze HJ (1983) Immunoelectron microscopic exploration of the Golgi complex. J Histochem Cytochem 31:1049.

Springall DR, Hacker GW, Grimelius L, Polak JM (1984) The potential of the immuno-silver staining method for paraffin sections. Histochemistry 81:603.

Spurr AR (1969) A low-viscosity resin embedding medium for electron microscopy. J Ultrastruct Res 26:31.

Stäubli W (1960) Nouvelle matière d'inclusion hydrosoluble pour la cytologie électronique. Compte-rendus Acad Sci (Paris) 250:1137.

Stäubli W (1963) A new embedding technique for electron microscopy, combining a water-soluble epoxy resin (Durcupan) with water-insoluble Araldite. J Cell Biol 16:197.

Steinbrecht RA (1982) Experiments on freeezing damage with freeze substitution using moth antennae as test objects. J Microsc (Oxford) 125:187.

Steinbrecht RA, Zierold K (1987) Cryotechniques in Biological Electron Microscopy. Springer-Verlag, Berlin.

Sternberger LA (1979) Immunocytochemistry, 2nd. edition. John Wiley and Sons, New York.

Stierhof Y-D, Schwarz H (1989) Labeling properties of sucrose-infiltrated cryosections. In: Albrecht RM, Ornberg RL (eds) Scanning Microscopy Supplement 3. Scanning Microscopy Intl., AMF O'Hare, Chicago, Illinois, pp 35-46.

Stierhof Y-D, Humbel B, Schwarz H (1991) Suitability of different silver enhancement methods applied to 1 nm colloidal gold particles: an immunoelectron microscopic study. J Electron Microsc Techn 17:336.

Stierhof Y-D, Schwarz H, Frank H (1986) Transverse sectioning of plastic-embedded immunolabeled cryosections: morphology and permeability to protein A-colloidal gold complexes. J Ultrastruct Mol Struct Res 97:187.

Stierhof Y-K, Humbel BM, Hermann R, Otten MT, Schwarz H (1992) Direct visualization and silver enhancement of ultra-small antibody-bound gold particles on immunolabeled ultrathin resin sections. Scann Microsc 6:1009.

Studer D, Michel M, Müller M (1989) High pressure freezing comes of age. In: Albrecht RM, Ornberg RL (eds) Scanning Microscopy Supplement 3. Scanning Microscopy Intl., AMF O'Hare, Chicago, Illinois, pp 253-269.

Taatjes DJ, Chen T-H, Ackerstrom B, Björck L, Carlemalm E, Roth J (1987) Streptococcal protein G-gold complex: comparison with staphylococcal protein A-gold complex for spot blotting and immunolabeling. Eur J Cell Biol 45:151.

Takada T, Yamamoto A, Omori K, Tashiro Y (1992) Quantitative immunogold localization of Na, K-ATPase along rat nephron. Histochem 98:183

Tobler M, Freiburghaus AU (1990) Occupational risks of (meth)acrylate compounds in embedding media for electron microscopy. J Microsc (Oxford) 160:291.

Tobler M, Freiburghaus AU (1991) Exceptional protective power of the 4H glove defeats occupational risks in electron microscopy. J Microsc (Oxford) 163:RP1.

Tobler M, Wüthrich B, Freiburghaus AU (1990) Contact dermatitis from acrylate and methacrylate compounds in Lowicryl embedding media for electron microscopy. Contact Dermatitis 23:96.

Tokuyasu KT (1973) A technique for ultracryotomy of cell suspensions and tissues. J Cell Biol 57:551.

Tokuyasu KT (1984) Immuno-cryoultramicrotomy. In: Polak JS, Varndell IM (eds) Immunolabelling for Electron Microscopy. Elsevier, Amsterdam, pp 71-82.

Tokuyasu KT (1986) Application of cryoultramicrotomy to immunocytochemistry. J Microsc (Oxford) 143:139.

Tokuyasu KT (1989) Use of poly(vinylpyrrolidone) and poly(vinyl alcohol) for cryoultramicrotomy. Histochem J 21:163.

Van Harreveld A, Crowell J (1964) Electron microscopy after rapid freezing on a metal surface and substitution fixation. Anat Rec 149:381.

VanWinkle WB (1991) Lectinocytochemical specificity in human eosinophils and neutrophils: a reexamination. J Histochem Cytochem 39:1157.

Villiger W (1991) Lowicryl resins. In: Hayat MA (ed) Colloidal Gold: Priciples, Methods, and Applications, Vol.3. Academic Press, San Diego, pp 59-71.

Villiger W, Bremer A (1990) Ultramicrotomy of biological objects: from the beginning to the present. J Struct Biol 104:178.

Voorhout W, van Genderen I, van Meer G, Geuze H (1991) Preservation and immunogold localization of lipids by freeze-substitution and low temperature embedding. In: Roomans G, Edelmann L (eds) Scanning Microscopy Supplement 5. Scanning Microscopy Intl., AMF O'Hare, Chicago, Illinois, pp S17-S25.

Weibull C (1986) Temperature rise in Lowicryl resins during polymerization by ultraviolet light. J Ultrastruct Mol Struct Res 97:207.

Weibull C, Christiansson A (1986) Extraction of proteins and membrane lipids during low temperature embedding of biological material for electron microscopy. J Microsc (Oxford) 142:79.

Weibull C, Christiansson A, Carlemalm E (1983) Extraction of membrane lipids during fixation, dehydration and embedding of Acholeplasma laidlawii cells for electron microscopy. J Microsc (Oxford) 129:201.

Weibull C, Villiger W, Carlemalm E (1984) Extraction of lipids during freeze-substitution of Acholeplasma laidlawii cells for electron microscopy. J Microsc (Oxford) 134:213.

Weibull C, Carlemalm E, Villiger W, Kellenberger E, Fakan J, Gautier A, Larsson C (1980) Low-temperature embedding procedures applied to chloroplasts. J Ultrastruct Res 73:233.

Westphal C, Böhme H, Frösch D (1985) Glycogen staining on sections of aqueous-embedded cyanobacteria and muscle. J Histochem Cytochem 33:1180.

Woldringh CL (1973) Effect of cations on the organization of the nucleoplasm in Escherichia coli prefixed with osmium tetroxide or glutaraldehyde. Cytobiologie 8:97.

Wroblewski J, Wroblewski R (1986) Why low temperature embedding for X-ray microanalytical investigations? A comparison of recently used preparation methods. J Microsc (Oxford) 142:351.

Wroblewski R, Wroblewski J, Wikström S-O, Anniko M (1990) A low temperature vacuum embedding procedure for X-ray microanalysis of biological specimens at subcellular level. Scann Microsc 4:787.

Wynford-Thomas D, Jasani B, Newman GR (1986) Immunohistochemical localisation of cell surface receptors using a novel method permitting simple, rapid and reliable LM/EM correlation. Histochem J 18:387.

Yoshimura N, Murachi T, Heath R, Kay J, Jasani B, Newman GR (1986) Immunogold electron microscopic localization of calpain I in skeletal muscle of rats. Cell Tissue Res 244:265.

Yoshitake S, Imagawa M, Ishikawa E, Niitsu Y, Urushizaki I, Nishiura M, Kanazawa R, Kurosaki H, Tachibana S, Nakazawa N, Ogawa H (1982) Mild and efficient conjugation of rabbit Fab' and horseradish peroxidase using a maleimide compound and its use for enzyme immunoassay. J Biochem 92:1413.

Zini N, Maraldi NM, Martelli AM, Antonucci A, Santi P, Mazzotti G, Rizzoli R, Manzoli FA (1989) Phospholipase C digestion induces the removal of nuclear RNA: a cytochemical quantitative study. Histochem J 21:491.

Index

Printing: Mercedesdruck, Berlin
Binding: Buchbinderei Lüderitz & Bauer, Berlin